# Successful Systems Engineering for Engineers and Managers

Norman B. Reilly

VNR VAN NOSTRAND REINHOLD
New York

Library of Congress Catalog Card Number 93-69
ISBN 0-442-01586-0



I(T)P Van Nostrand Reinhold is an International Thomson Publishing Company.
ITP logo is a trademark under license.

Printed in the United States of America.

Van Nostrand Reinhold
115 Fifth Avenue
New York, New York 10003

International Thomson Publishing
Berkshire House
168-173 High Holborn
London WC1V7AA, England

Thomas Nelson Australia
102 Dodds Street
South Melbourne 3205
Victoria, Australia

Nelson Canada
1120 Birchmount Road
Scarborough, Ontario
M1K 5G4, Canada

16 15 14 13 12 11 10 9 8 7 6 5 4 3 2 1

**Library of Congress Cataloging-in-Publication Data**

Reilly, Norman B.
Successful systems engineering for engineers and managers / Norman B. Reilly.
p. cm.
ISBN 0-442-01586-0
1. Systems engineering. I. Title.
TA168.R375 1992
620'.001'1—dc20 93-69
CIP

# Contents

# Acknowledgments

The author of any book, large or small, influential or fleeting, is necessarily indebted to a lot of people. Among those owed a significant debt is the host of teachers and professors who so willingly demonstrate their faith throughout our years of learning. Whether we write books or not, we do well to honor them.

More recently, I am clearly indebted to Glen Garrison, Chris Carl, Dave Werntz, Steve Loyola, and Dick Morris, all of the California Institute of Technology's Jet Propulsion Laboratory. They have helped me through failure and success to begin to grasp the system perspective. Garrison is a superb engineer as well as a gifted and patient teacher. Carl has to be one of the finest systems engineers around. The young geniuses Werntz and Loyola were of great assistance in improving sections of this book. Morris has a habit of asking highly structured questions about topics you think you know something about. When he is finished, you realize that you don't really know very much at all. He has done this incessantly for me over the years. Socrates would be proud of him.

In addition, I wish to thank the highly professional assistance and support given me by the extremely capable staff at Van Nostrand Reinhold. A colleague of mine neatly and aptly described the people at VNR as being "on the cutting edge."

It is also a special pleasure for me to express my appreciation for the editorial assistance, patience, and enduring encouragement of Patricia Donnelly.

# Why This Book?

There is a common consensus that the need for systems engineering arose in this century as a direct result of technological complexity. In all fairness, however, engineers of previous eras faced with the design, integration, logistics support, and configuration management of any complex enterprise involving the interaction of multiple disciplines would certainly have had to work their way through a myriad of technical and personnel issues in order to unite diverse technologies and interests into a useful whole. The words may have been different, but the rules for success are the same.

While systems engineering may not be new, the consensus is still partially right. Certainly, more people are constructing extremely complex hardware and software systems than ever before. Many of today's technical organizations include whole divisions devoted to this discipline. Within such divisions one commonly finds an abundance of sections and groups with the words "systems analysis" or "systems engineering" in their titles.

Despite this proliferation, the road to discovering what systems engineering is, when it starts, when it ends, and how one does it for different classes of projects has been a difficult and empirical one. While many successful complex systems have been realized, our achievements are still paled by an uncomfortable number of major system efforts undertaken over the past 40 years (many very recently) that were simply abandoned in mid-course or failed to meet user requirements upon delivery. Failures to implement the simplest and basic tenets of the systems engineering process remain disturbingly prevalent.

Throughout this period, a number of major conditions have been evident:

1. The discipline of systems engineering has gone through an orderly evolution in a constant effort to devise ways of increasing the probability of success when developing large, complex systems. This evolution is not widely understood.

2. Progress in the field has typically been documented in professional journals, compilations, and internal organizational publications. The knowledge is highly distributed.
3. While most organizations have a number of qualified systems engineers, there are simply not enough of them. Those that are proficient are in high demand and typically too busy to train their more needy colleagues. In the meantime, because the process itself is so ill-defined within the technical culture, the title of "systems engineer" is liberally distributed by middle and upper managers to the partially, and often thoroughly, uninitiated. A result is that the good systems engineers are survivors. They are good because they have learned what has to be done through a series of very difficult experiences, without the supporting benefit of access to a thorough structured methodology for the execution of all facets of systems engineering.
4. There is no single, widely accepted definition of what systems engineering is—any two "systems engineers" are likely to provide different definitions.
5. Existing material on the subject emphasizes the technical side. Clearly, technical competence is a necessary condition for success as a systems engineer, but it is not sufficient. Systems engineers must also exhibit formidable skills in their human relations in order to motivate diverse human interests toward a productive consensus. Things don't do things—people do.
6. Experience dictates that the implementation of modern systems engineering practices often requires new perspectives in traditional management structures and approaches.

The purpose of this book is to help improve these conditions in at least two ways. The first is to compile a single, comprehensive resource to serve these needs by defining the scope and responsibility of modern systems engineering for managers, engineers, and students of the subject. The second is to promote the standardization of systems engineering by providing a structured methodology for the successful execution of the systems engineering process.

# 1

# Defining Systems Engineering and Systems

There can be little doubt that efficient communication in any of the sciences demands rigorous use of terminology. Still, there are so many words that systems engineers use routinely, over and over, day after day, that their repeated usage alone lulls us into a perception that a common understanding exists. But in many cases it doesn't. This is especially true of words that seem easy to understand, whose meaning and scope we assume we know and seldom bother to track down. In reality, words such as *requirements, management, top-down, logistics support, work breakdown structure,* and *systems engineering* have surprisingly diverse meanings to most people who routinely use them. While we don't seem to have the same problem with words like *mass, acceleration,* or *sphere,* there is evidently a very real need to decide upon the precise meanings of words used by "systems engineers."

Having said this, the following deceptively simple words shall be used to define *systems engineering.* This set of words shall then be dissected in the interest of common understanding.

## DEFINITION OF SYSTEMS ENGINEERING

Systems engineering is the systematic application of proven standards, procedures, and tools to the technical organization, control, and establishment of:

- System requirements;
- System design;
- System management;
- System fabrication;

- System integration;
- System testing; and
- System logistics support.

## The Meaning of the Meaning

This section will clarify the meaning of each significant term used in the above definition.

**Systematic application**—This term implies that there is a definite generic sequence of activities that must be consistently followed. This systems engineering process is treated in Chapter 3. The early stages of the process are very critical, since they determine the structure, sequence, and completeness of all work that is to follow. In these early stages, errors of omission are highly prevalent and are the most devastating. However, the complete sequence must not be forgotten or abandoned, but must be applied systematically throughout the entire systems engineering process.

**Proven standards**—These are standards, such as those published by the IEEE, the military, and other groups, that have been issued, widely used, and revised through practical experience. They also include standards produced by individual organizations that have proven to be of value. Good standards are of significant worth, even when not formally imposed on a project, because they help to assure that all aspects of an issue have been considered. Incomplete reinvention is not necessary. Virtually every "How should I do this?" question faced by a systems engineer has been faced before by others. The library of standards proven to work represents a collection of solutions. However, standards should be looked upon as guidelines that complement and complete the thinking process. Even when formally imposed, they remain negotiable in accordance with agreed-upon project priorities and policies.

**Procedures**—Refers to the actual set of procedures that must be "systematically applied." Examples of procedures are the definition of user needs, determination of system constraints, development of prioritized competing design characteristics, work breakdown structures, design team formation and use, configuration management plans, integrated logistic support plans, test plans and procedures, functional requirements, options analysis, specifications, design and fabrication methodologies, scheduling, cost accounting, and performance measurement techniques.

**Tools**—The set of tools used to implement procedures. Examples of implementation tools are unit folders, structured analysis and design, object-oriented programming, design languages, computer-aided design, queueing theory, simulation, breadboarding, prototyping, margin management, traceability, pert charts, design reviews, earned value concepts, walk-throughs, and configuration management techniques. Tools have one simple purpose—to provide *visibility* to the systems engineer, to implementors, and to management. Tools provide visibility because they enhance everyone's understanding of what has to be done and whether it is being done on time and within budget.

**Technical organization**—Refers to the allocation of technical responsibilities to project management, project engineering, and systems engineering, and to subsystem and specialty cognizant engineers. Many organizations are simply not organized to do systems engineering. Committing to a systems engineering role in a project organization where technical responsibilities are not clearly understood continues to be a classic cause of failure (see Chapter 15).

**Control and execution**—The act of systematic application of proven standards, procedures, and tools. Control cannot be achieved without visibility tools and interpersonal skills.

**System requirements**—A subset of global user-stated needs adapted by mutual consent of implementor and client at the *system requirements review* milestone, stating *what* the system must do. All system requirements are measurable (see Chapter 10).

**System design**—The process of determining *how* system requirements shall be met. System design activities involve varied levels of detail, from top-level system partitioning to detailed concept verification through modeling, breadboarding, prototyping, and trade-off analysis, as required (see Chapter 12). The major product of system design is the complete specification.

**System management**—Consists in part of the four major elements of configuration management:

1. Identification. (What are the elements that must be controlled?)
2. Configuration control. (What are the mechanisms and procedures for controlling changes?)
3. Auditing. (What is the current status of the system?)
4. Status reporting. (To whom, when, and how is this information reported?)

Configuration management is often viewed as a burden, something that upper management forces upon us. Far from a burden, it is a major tool for maintaining visibility as the systems engineering process matures.

Other major system management activities include the development and execution of the *systems engineering management plan* (see Chapter 6), leadership of the *system design team* (see Chapter 7), and the mature use of *management guidelines* (see Chapter 15).

**System fabrication**—The act of cutting metal and generating code—that is, the physical realization of the system. This may or may not involve production prototyping or large-scale production. Fabrication does not begin until appropriate *critical design reviews* have taken place. System fabrication includes unit testing.

**System integration**—The act of planning for, designing, and executing the sequencing of subsystem interconnection into a total system accompanied by appropriate system integration testing (see Chapter 14).

**System testing**—The activities involved in developing test plans and procedures for system integration testing, system testing, and system acceptance testing; also, those activities involved in their execution (see Chapter 14).

**System integrated logistics support**—The act of designing and providing for system supply support, required test equipment, transportation and handling, technical data packages, facilities, personnel and training, and a maintenance plan. There are at least two separate ILS systems required—one to support development and another to support operations (see Chapter 11).

## WHAT IS A SYSTEM?

The Army field manual on systems engineering provides a useful definition of the word "system" as follows:

> A composite of equipment, skills and techniques capable of performing and/or supporting an operational role. A complete system includes related facilities, equipment, material, services, software, technical data and personnel required for its operation and support to the degree that it can be considered a self-sufficient unit in its intended operational and/or support environment. The system is employed operationally and supported logistically.[1]

Note that this definition includes users (skills and techniques) in addition to the physical system and its logistics support.

The use of the term "self-sufficient" should not be interpreted to suggest that systems do not have inputs and outputs. They do. Rather, the term means that systems have well-defined functional boundaries. The boundary may contain a simple function or it may contain many functions.

The military definition is oriented around operational roles. In a broader sense, examples of systems are the universe, a neuron, an F-16 fighter, an automobile, a tree, a star, a locomotive, a rabbit, the weather, the Third Army, a language, a publishing house, a clinical laboratory, an inertial navigation set, a manufacturing plant, a marketing organization, and a toadstool.

Clearly, a system can be a part of a larger system. You may consider the sun as a system in itself or as a subsystem if your system perspective (definition) is the solar system.

An early step in the systems engineering process is system and subsystem definition. This is done by using tools for determining the *work breakdown structure* and *product breakdown structure* to meet specified requirements.

While there can be, and are, a myriad of "systems," all systems consist of an interdependent group of items (subsystems) that perform together as a functional unit. The functional boundary is what determines the system definition in any given application of the concept.

System definition is a crucial and interesting exercise. It involves applying a seemingly indistinct concept to a highly specific environment. When the process is completed, the system boundary must be defined with momentous precision and distinguished rigor.

**References**

1. *United States Army Field Manual—Systems Engineering*, FM 770–78.

# 2

# The Evolution of Systems Engineering

The discipline of systems engineering is a dynamic one. Over the past 40 years it has methodically evolved through the *staircase, early prototype, spiral,* and *rapid development* models. The evolution is characterized by successively deeper penetrations of prototyping concepts into the traditional staircase paradigm.

Early prototyping penetrates to the requirements level. The spiral model penetrates to the design level. Rapid development, the most recent innovation, engulfs the entire process, resulting in repeated operational deliveries. Each paradigm continues to have its useful niche. Awareness and proper application of this evolutionary perspective can be of significant value to project and systems engineers. Selecting the wrong model for a given task will increase the probability of turmoil and, as history warns us, even failure. There is a need for continued education and acceptance of the newer approaches within the technical management and systems engineering communities.

This chapter presents an overview of this pattern, suggests specific application examples, and offers comments on our need to continue to adapt as systems engineers and managers.

## A SIMPLE AND QUICK LOOK BACK

In the beginning the paradigm was Build it, Test it, Fix it. During the initial growth of access to Von Neumann resources in the 1940s, this was, in fact, what software engineering was mostly about. It has also been the hallmark of much of our hardware engineering for centuries. For smaller systems the approach worked quite well. It may still be an acceptable approach when the developer is also the user and there is little need for external communication. With systems of any reasonable complexity, however, the approach often compromises adherence to user requirements, adequate testing, acceptable documentation, and maintainability, and results in a horde of other shortcomings. The approach also typically generated systems that, when finally "fixed," tended to exhibit ungraceful architectures.

With the advent of "big" systems, better ideas emerged. The building of the SAGE (Semi-Automated Ground Environment) system at TRW in the mid 1950s was among the first recognized attempts to do business in a more structured manner. The philosophy was to execute discreet and controlled stages of system development that consisted of the orderly determination of requirements, specifications, and design, followed by fabrication, test, and the performance of acceptance testing prior to operations. Figure 2-1 depicts the procedure as the staircase model along with the other systems engineering paradigms discussed in this paper. In this scheme, each step is completed before the next is begun. At the time it was an innovative, bold, and constructive idea fostered by the need for control and visibility.

However, giant step that it was, the idea still had drawbacks. It was soon discovered that any of the discrete stages in the staircase were almost always

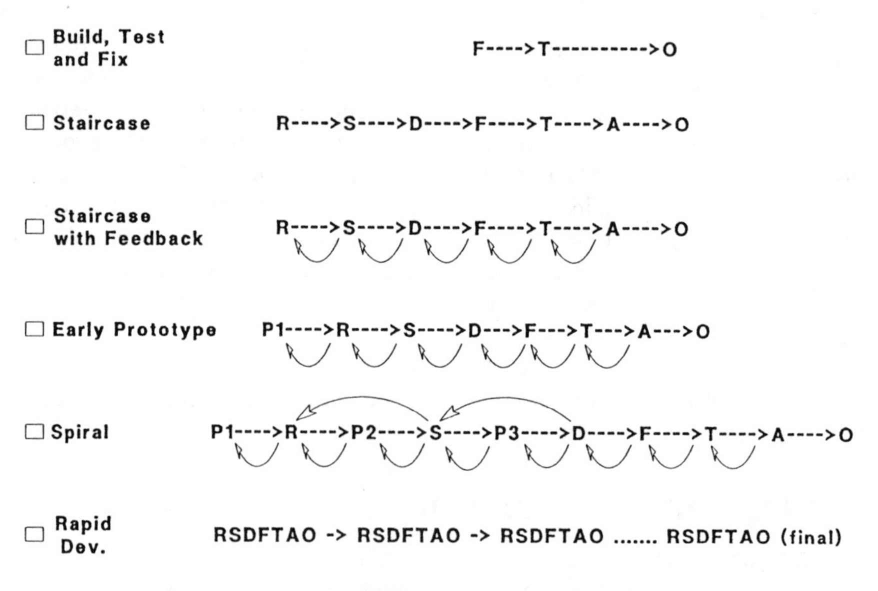


**FIGURE 2-1.** Systems engineering process paradigms.

modified by further insights gained in executing the next step. The realization that changes resulting from such insights were of value and even necessary for success led to the next evolutionary modification—the *staircase with feedback*. This approach recognized that one needed to at least partially explore a subsequent stage in order to gain full understanding of the present stage. More simply put, humans rarely organize an increasingly complex series of events without gaining insight down the road on how to start out better the next time. The addition of appropriate freedom to feed such knowledge back to a previous step in the staircase model is the substance of the staircase with feedback idea. The new model visualized a block of effort that moved forward in time, encompassing both the present stage and the next stage. While this paradigm upgrade did not recognize a need for concurrent feedback across multiple stages, it clearly helped to introduce a reality of human behavior into the process.

The staircase with feedback paradigm has had a powerful and lasting influence on the way we do systems engineering. It gave birth to an entire lore of control mechanisms to insure the maintenance of its orderly flow. These new disciplines and techniques included classic configuration management, standardization of reviews, formalization of the system design team, third-party quality control, and test teams, to mention a few. In time, it also gave rise to a host of ancillary techniques designed to maintain the integrity of the staircase with feedback concept when it more than occasionally struggled or faltered. These additional tools included structured analysis, structured design, structured programming, object-oriented programming, program design languages, unit development folders, computer-aided systems engineering tools, and other similar techniques designed to maintain control and visibility. Throughout this period, the staircase with feedback paradigm remained solidly entrenched. The belief persisted that the model could inherently be successful when accompanied by adequate controls and sound management.

The staircase with feedback model remains valid if certain conditions are met. Perhaps the most important condition for its success is that customers completely understand their requirements at the beginning. The next step in the evolution, early prototyping, found its impetus when it was recognized that this is not always the case.

Formal rumblings with the staircase feedback fashion appeared in the early 1980s. A Government Accounting Office (GAO) survey, reported in October 1980, disclosed that only 2 percent of the systems surveyed were used as delivered [5]. Fully 47 percent were delivered but never used, and 29 percent were never delivered. The remaining 22 percent were used only after moderate-to-extensive rework. Cost and schedule overruns accompanied many of these failures.

Evidently, users were simply not getting what they thought they were going to get. The problem was perceived to be due largely to a basic failure to meet requirements as originally stated. A realization emerged that there are at least two conditions under which the user is unable to state requirements in sufficient detail to realize success. The first is when users cannot state exact requirements, but can recognize them once they are exposed to them. The second condition arises when no one really knows what the requirements are in detail until the system has gone through a number of actual operational iterations. An example of the first condition might be a new end-user command and control system where the user may not know the exact desired information content or the form and format for screen presentation in advance. An example of the second condition is the space station (i.e., what is the mission of a space station with a life of, say, 20 years going to be 15 years from now?). Not a lot was done immediately about the second problem. But the first problem was attacked throughout the 1980s with expanded use of early prototyping.

The concept of early prototyping accepts that the user may not always know what the exact requirements for a system are. In this scheme, user requirements would be drawn out by constructing a set of "best guess" requirements at the outset. The set would then be rapidly prototyped and tried out in a mock-up of the user environment. Appropriate modifications would be made until the user was able to clearly identify the desired system behavior. When requirements were finally determined in this manner, the systems engineer could then revert to the staircase with feedback model for all the remaining stages.

Early prototyping was widely extolled and spread rapidly and comfortably into the process of systems engineering. Not only did individual projects embrace the new paradigm, but soon whole laboratories devoted to early prototyping appeared as organizational resources that assisted projects in accurately determining fuzzy requirements.

Despite these advances, in 1987 the Defense Science Board Task Force Report on Military Software concluded that traditional software development paradigms were actually hindering effective software development [6]. The Board was not alone in its observations. Significant system "failures" were still being observed, and software engineers were responding. Among the responses was a new and significant evolutionary concept in software systems engineering that began to appear in the literature in the mid-to-late 1980s [1, 2]. It was called the spiral model.

The spiral model, which also applies to systems comprised largely of hardware, extends the prototype concept beyond a single, early prototype to a series of three prototypes. The paradigm is depicted linearly in Figure 2-1 for simplicity, but it can be visualized as going around three times before reverting to the staircase with feedback paradigm, upon the final establishment of a design. The first prototype aims at developing system and operational requirements, the second aims at developing specifications, and the third results in the hard establishment of the design. The spiral model may incorporate more or less than three prototypes, depending upon the need. Each spiral involves extensive risk analysis, the use of simulations, models and/or benchmarks, and user interaction, as required. While the impetus for this evolution came from software systems engineering, the concept is clearly applicable to the larger arena of systems engineering in general. The approach basically recognizes that uncertainty typically exists beyond the requirements stage, and deals with this issue by reaching deeper into the conventional model with a series of penetrating prototypes.

The spiral model exhibits a very distinct and important advantage over the earlier paradigms in that it provides for unforseen change deeper into the process. It provides a mechanism for dealing with uncertainty, as well as the capacity to accommodate trade-offs between implementation and requirements. In the spiral model, requirements are not really frozen until design, cost, and risk issues are adequately addressed.

Further, the older paradigms tended to separate the user from the developer early in the process, each to go their merry way until perhaps a not-so-merry delivery. Clearly, one lesson to be learned from the observed progress in paradigms is that, if the customer is an intimate partner in trade-off analyses, the probability of success is greatly enhanced. The tendency for engineers to design systems for engineers as opposed to designing them for end users is thus somewhat reduced.

Finally, proponents of the spiral model suggest that, in conditions where uncertainty is not so threatening, the approach can revert at anytime to the staircase with feedback. In this sense, the earlier models can be viewed as subsets of the spiral model.

Use of the model, however, does require personnel experienced in risk assessment, as well as experienced managers. It also requires a further modification to our classical ideas about configuration management. Use of the model impacts the concept of frozen baselines at the requirements, design, and implementation levels. Clearly, the conventional review process involving user requirements reviews, preliminary design reviews, and so on, must now be modified to recognize the volatility of solution choices and their potential impacts on requirements throughout the spiraling process.

While the spiral model represents a significant advance in the evolution of systems engineering, a further evolutionary concept designed to deal with the shortcomings of the staircase with feedback approach has recently begun to emerge. This newest concept addresses the issue of what happens when a substantial portion of true requirements are not really understood until the system is actually operational? When this occurs under the conventional development paradigms, the resulting system is typically discarded because the required changes outstrip the ability of the design to easily respond to the newly discovered requirements or because the prospect of reinitiating a staircase with feedback fix (i.e., starting all over again) is simply too costly or time consuming. Evidently, these are the systems that gave much of the fodder to the previously referenced GAO study.

Clearly, systems like this exist—more than we may wish. And here, in the early 1990s, yet a new and imaginative innovation is extending the prototype trend to include repetitive iterations of the entire traditional end-to-end systems engineering process. It is called rapid development [4]. Use of the word rapid does not mean that the complete system is necessarily developed rapidly. Rather, it means that the final system is brought to fruition through a series of deliveries, each of which is some 10 to 18 months apart. The effort involved in each delivery encompasses the entire staircase process, and each delivery represents a further hierarchical accomplishment on the road to final detail. These mini-waterfalls allow successive injections into the actual operational environment and in so doing absolutely insure continual and intense interaction with the user.

But this does not all happen without controls. In fact, the controls are perhaps more stringent than in the more conventional processes. Each delivery or repeated phase involves definite commitments, such as user reviews, configuration control reviews, system and subsystem functional freezes, designs, fabrications, test plans, pre-ship reviews, and transfer reviews. These are similar, in spirit, to milestones found in the conventional configuration management process. What is substantially different, however, and inherent to the rapid development concept is that functional and design requirements are free to evolve in each phase, based on experience gained from the previous phase delivery. This is decidedly different from the staircase with feedback paradigm, where requirements must be finalized at the beginning. Requirements are not actually finalized using the rapid development approach until the last phase ends. They are, however, almost guaranteed to be correct. Rapid development thus appears to complete the penetration of prototyping concepts into the systems engineering process to include complete repetitions of the entire conventional process.

The steps carried out in each of the paradigms discussed are typically carried out sequentially. During each step, however, a vision for all factors that may impact design must be maintained. A more recent development aimed at addressing the need for that vision is embodied in the concepts of concurrent engineering. Concurrent engineering is defined as:

> A systematic approach to creating a product design that considers all elements of the product life cycle from conception through disposal. In so doing, concurrent engineering simultaneously defines the product, its manufacturing processes and all of the related life cycle processes, such as logistics support. Concurrent engineering is not the arbitrary elimination of a phase of the existing, sequential feed-forward engineering process, but rather the co-design of all downstream processes toward a more all-encompassing, cost effective optimum.[3]

Note that concurrent engineering concepts do not replace the paradigms themselves—that is, consideration of all factors that impact design has always been the responsibility of the system design team in the execution of any of the paradigms. What the ascent of concurrent engineering has done, however, is to emphasize the impacts on design that suppliers, the development process itself, production, and customer satisfaction can and should have. To this extent, concurrent engineering is directly akin to concepts of modern quality. These emphases have been positive factors in the evolution of systems engineering in that they have heightened the awareness of systems engineers as to the complete breadth of their responsibilities.

In practice, the scope of concurrent engineering can vary widely, depending on the product. A basic tenet of the concept calls for simultaneous consideration of product development and such issues as design impacts of manufacturing and logistics support. For this reason, the term *simultaneous engineering* is also often used. The notion may also include a certain amount of parallelism in development. An example of its simplest form may be the use of inherited equipment (an instrument, a power plant, etc.) that will require minor modifications for integration into a new system. Such modifications can often be designed prior to the rest of the system being designed. Thus, preliminary and detailed designs may take place on different parts of the system in parallel. In this setting, inheritance is considered a design constraint (see Chapter 3).

In its more generic form, concurrent engineering may involve consideration of requirements and design impacts from marketing, manufacturing, finance, test engineering, suppliers, safety, dealerships, customer service, and so on, as well as all facets of integrated logistics support for both development and operations.

In any case, the system design team is the focal point for concurrency (see Chapter 7).

The overall pattern of evolution we now observe was, of course, not a planned one. The attempts of systems engineers to devise schemes to avoid failure did not take place with the benefit of an overall vision or direction. Rather, they were initiated by creative individuals attacking, step by step, what was perceived to be the most pressing contributor to the continuing occurrence of failures. The pattern is only seen in retrospect.

This view of our past is useful in that it is now evident that the selection of the proper paradigm or an agreed-to modification is of paramount importance in increasing the probability of meeting the user's real needs.

There is a very important reason why the newer paradigms have been used with some success in situations where the older ones have failed. The reason is that the newer models tend to benefit from user feedback further into the implementation process, while the older paradigms inherently assume that the implementors can essentially be cut loose in quasi-isolation as soon as the requirements baseline is frozen. The latter strategy does not always work. It is quite likely that any successes realized by use of the newer systems engineering models is solely due to this reality. Perhaps the most important observation to be made from all of the history noted is that the user, in any paradigm, must be consistently and intimately involved throughout the entire design and implementation process. It is not sufficient for the user to simply attend high-level reviews. It is also clear that management must be more skilled when dealing with the more recent paradigms.

These are substantial lessons to learn from our past. While the development of the newer models has clearly helped, we must not be lulled into believing our work is done. We need only note that "big system" failures continue to occur at uncomfortable frequencies. Our hope for the future lies in the organization and understanding of our past.

Table 2-1 summarizes the historic motivations for and time frames of development for each of the paradigms discussed.

Table 2-2 lists a number of application domain characteristics and examples for each of the paradigms. The examples are not absolute, but are typical of their categories. It is the responsibility of technical management and systems engineering to carefully assess the applicability of the various paradigms to the particular tasks at hand, and then to make the user or qualified user advocates a permanent part of the team.

## COMMENTARY

The evolution we have seen has happened for a reason. The reason is tied to a historic attempt on the part of creative systems engineers to identify problems associated with existing process models and to improve upon them. The improvements have been real, in that each of the evolving paradigms has accumulated its own cluster of successes. The implication for systems engineers is that it is of considerable importance to map the right paradigm or a modification of it into and onto the problem domain at the outset. This can

**TABLE 2-1 The Evolution of Systems Engineering Paradigms**

| Paradigm | Motivation/Time Frame |
| --- | --- |
| Build, test, fix | Lack of experience during the 1940s during the rise of new computer-based systems. The approach compromised requirements, testing, documentation, and maintainability. |
| Staircase | Realization that specific phasing was required, consisting of controlled development of requirements, specifications, design, fabrication, test, and integration. Characterized by the TRW development of the SAGE system in the 1950s. |
| Staircase with feedback | Recognition in the 1960s and 1970s that phases could not be isolated—that feedback between each successive phase could greatly enhance success. |
| Early prototype | Recognition in the late 1970s and early 1980s that the customer did not always know what was needed. Early prototyping developed to define requirements. |
| Spiral | Recognition during the 1980s that implementation options were not always clear at the outset. The spiral model, developed at TRW, extended prototyping to the design phase and added risk analysis. |
| Rapid development | Recognition in the late 1980s that required system behavior could not always be determined until some operational experience took place. JPL formalizes the repeated system delivery concept. |

only be accomplished by understanding the patterns of our past and the reasons for their emergence. Clearly, imposing the staircase with feedback paradigm on a system that calls for rapid development would be and has been devastating. In fact, experience suggests that projects tailored to the wrong paradigms tend to gravitate toward the right paradigms by virtue of their own nature. That is, iterations are forced upon the developer by the user's experience. This need for accommodation of change occurs whether it is planned

**TABLE 2-2 Application Domains for Systems Engineering Paradigms**

| Paradigm | Characteristics | Application examples |
|---|---|---|
| Build, test, fix | Simple system, single developer, one or few users, minimum need for documentation | Home built telescope, personal application software |
| Staircase | Duplication of an existing system with straight-forward and minor modifications | Build of F-15 aircraft for export with minor avionics changes |
| Staircase with feedback | Requirements are clear and implementation technologies are bounded | Exploration spacecraft, locomotives, instruments |
| Early prototype | Complete requirements unknown but can be recognized by users, technologies mature | $C^3I$ systems, complex MIS systems, automobiles |
| Spiral | Trade-offs between requirements, cost, risk, and implementation strategies unclear | Same as early prototype but uncertainty is greater |
| Rapid development | Complex, new systems, never attempted before, requirements and final design destined to evolve with operational experience | Space station, novel command and control systems, complex biomedical systems |

for or not. When unrecognized, it often leads to both technical and programmatic chaos, if not outright failure to meet commitments.

We may further observe that as systems engineering evolves—and it is clearly doing just that—our concepts of configuration management must also evolve to serve it well. Perhaps one of the reasons that contemporary technical management can be so difficult at times is that the discipline of systems engineering continues to create innovative life cycle ideas without sufficient attending change in supporting configuration management concepts. This observation can be duplicated for reviews, documentation, quality assurance, margin management, logistics support, risk analysis, testing, acceptance criteria, and all the other familiar mechanisms that were spawned by the staircase with feedback model.

It is likely that the formalization of management and control techniques will always lag behind the innovation of new systems engineering processes to some extent. There is a current need within the systems engineering community to further develop this formalization for the newer paradigms. There is also a need to profit from the structure of our past by increasing the exposure of practicing systems engineers to the process options available to them.

## CONCLUSIONS

The following significant conclusions can be drawn from the evolutionary view presented in this chapter:

1. The discipline of systems engineering is dynamic. It has evolved through experience and will most likely continue to evolve.
2. The evolution of systems engineering to date has followed a pattern. The pattern has focused on the increased migration of the penetrating prototype into the conventional process.
3. Experience dictates that successful system development requires intensive user interactions throughout the entire development life cycle process.
4. As the element of uncertainty increases, the systems engineer should turn to process paradigms that have most recently appeared on the evolutionary scale.
5. An early routine step in the systems engineering process should be the selection of an appropriate life cycle paradigm or an agreed-to modification.
6. Experience dictates that quality and user satisfaction is greatly enhanced through the use of concurrent engineering principles that consider all factors that can impact development, production, and operational use for any of the development paradigms.
7. There is a need for further development and standardization of management and control techniques for the newer paradigms. The next contributions to the evolution of systems engineering should focus on this need.
8. There is a need for continued education and acceptance of the newer systems engineering paradigms, with regard to their use and applicability within the technical management community.

**References**

1. Boehm, B. W. 1988. A spiral model of software development and enhancement. *IEEE Computer Journal*, May 1988.

2. Boehm, B. W., and F. C. Belz. 1988. Applying process programming to the spiral model. Proceedings of the Fourth Software Process Workshop, IEEE, May 1988.
3. Concurrent engineering in system engineering. 1992. Second International Symposium, National Council on Systems Engineering, Seattle, Washington, July 1992.
4. Spuck, W. H. 1991. *Management of Rapid Development Projects*. Jet Propulsion Laboratory, D-8415, April 8, 1991.
5. *Contracting for Computer Software Development—Serious Problems Require Management Attention to Avoid Wasting Additional Millions*. 1979. Report by the Comptroller General to the Congress of the United States, FGMSD-80-4, November 9, 1979.
6. Brooks, F. P. 1989. *Defense Science Board Task Force Report on Military Software*. Office of the Under Secretary of Defense, September 1989.

# 3

# Systems Engineering Process Overview

Figures 3-1 through 3-4 provide an overview of the systems engineer's perspective throughout the complete process of system realization, from *definition of user needs* to the achievement of the final *product baseline*. The process depicted follows the staircase with feedback model. When the spiral model is used, emphasis on the use of prototyping and risk analysis is incorporated during the requirements definition and design phases. The rapid development paradigm basically includes a complete iteration of the staircase model for each successive delivery.

Major processes are depicted in boxes. The sequence of the processes is depicted by connections between the boxes. The overall process, however, is not strictly sequential. That is, while the concentration of effort deals with a single box at a time and moves downward through the paradigm, there is typically considerable iteration with neighboring boxes above and below.

This chapter presents a brief description of the function(s) represented by each box. References to chapters containing more detail are provided.

The overall process is strictly segmented by three major review milestones. Systems engineering, when correctly executed, is a *front-loaded* enterprise. The most crucial work is accomplished prior to the system requirements review. The system requirements review is primarily devoted to proving to the sponsor that the problem (what the system must do) is understood, requirements are complete and documented, and a technically feasible solution exists. This is an extremely pivotal period in which understandings must be thorough and complete. Errors of omission in the process at this point plant the seeds for distant and complex repercussions—none positive.

Progression to the *preliminary design review* is also of significant importance, but much of the peril can be contained if the previous course to the

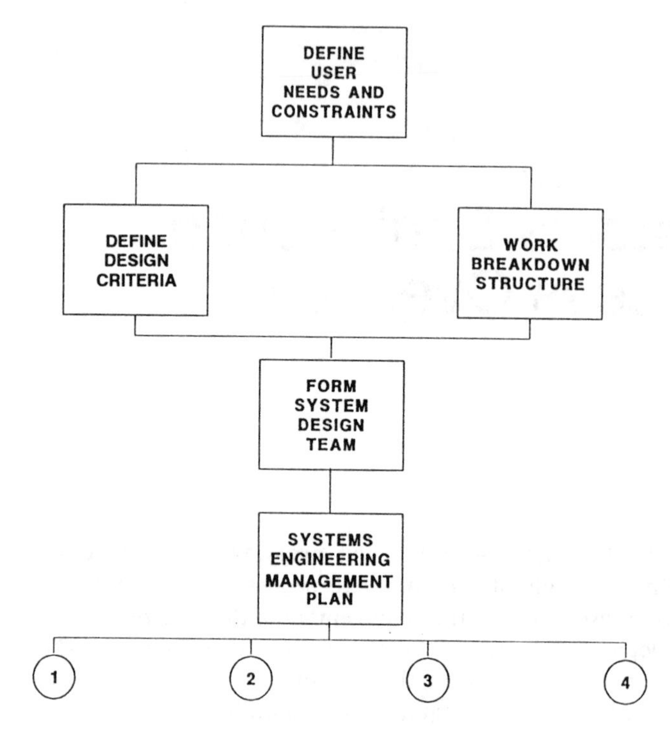


**FIGURE 3-1.** Systems engineering process overview.

system requirements review was properly executed. The bowling ball analogy is well suited. At the preliminary design review you exhibit the approach to "how" you intend to implement the "what" presented at the system requirements review and seek approval to carry out a detailed design. Note in Figure 3-2 that prior to the preliminary design review, a system design review and/or a *system specification review* may or may not be desirable. Including these extra reviews is largely determined by the degree of confidence experienced by project management and the systems engineer at this point in the overall process.

Having accomplished these milestones, the avenue to critical design review should involve less ambiguity. At the critical design review you seek approval to cut metal and code—that is, begin actual fabrication. There may be a number of critical design reviews, some for different aspects of the system, depending on the level of implementation confidence.

Again, most of the thought and hard work that goes into the system engineering process must occur at the beginning. The process is front loaded only when it is done correctly. If the necessary diligence to completeness is

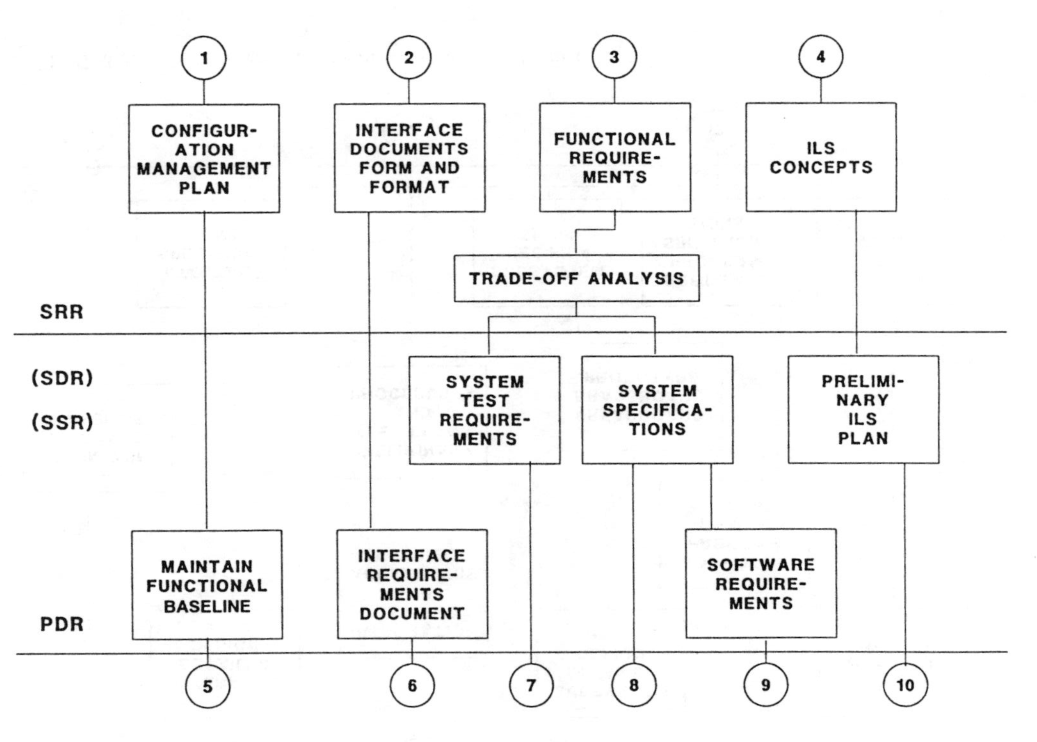


**FIGURE 3-2.** Systems engineering process overview (cont.).

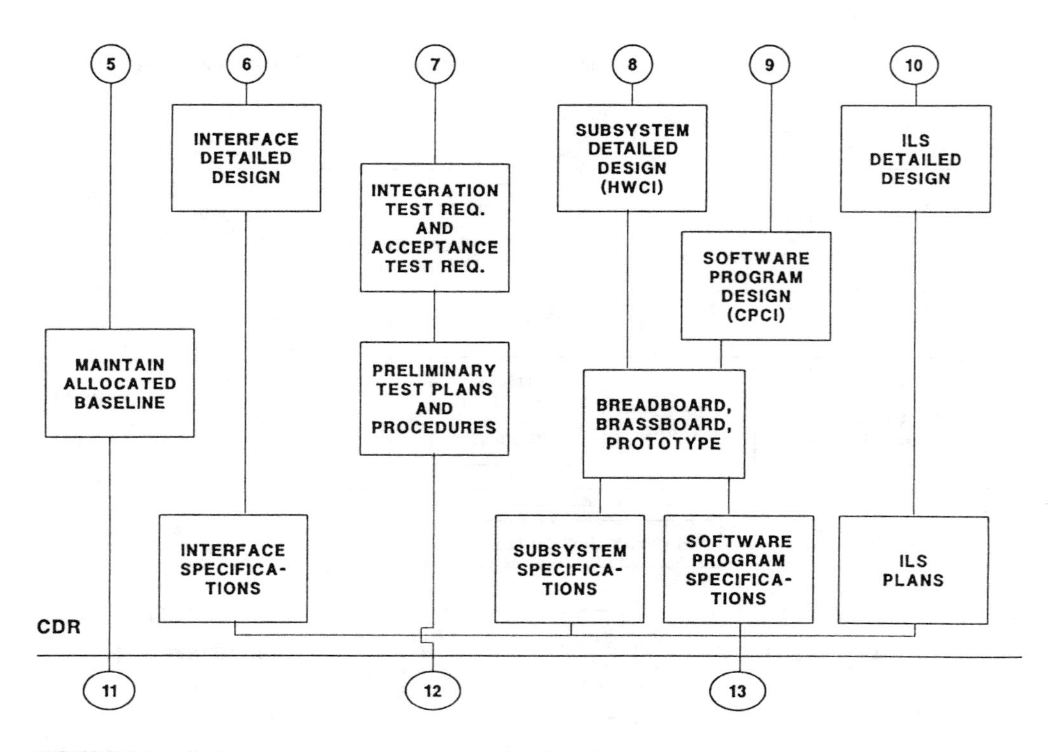


**FIGURE 3-3.** Systems engineering process overview (cont.).

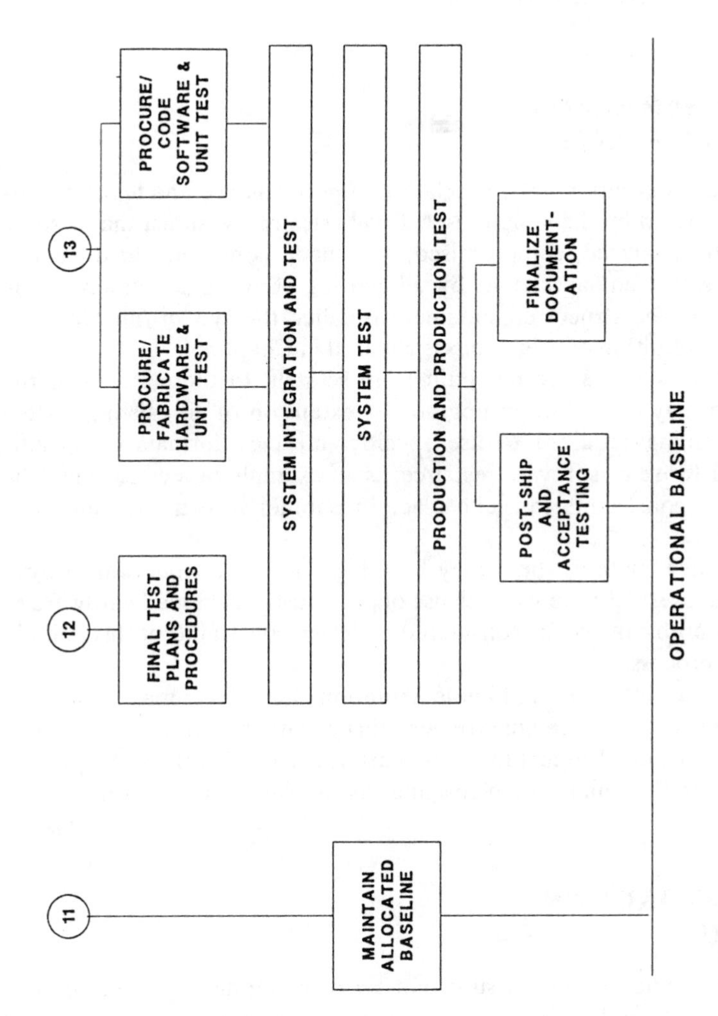


**FIGURE 3-4.** Systems engineering process overview (cont.).

not applied up front, the process becomes *back loaded*—reactive, corrective, recursive, full of patches and design changes, subject to chaos, and generally demoralizing.

The following paragraphs provide top-level descriptions for each element presented in the overview paradigm.

## DEFINE USER NEEDS AND CONSTRAINTS

The user need definition is a compilation of performance and logistic characteristics desired by the system user. It is basically a wish list that is to be subsequently analyzed, compromised, and made consistent, leading to a structured statement agreed to by all parties of what is actually to be implemented. The structured statement is called the *system functional requirement.* Definition of user needs is covered in Chapter 4.

Very often, systems are not built from scratch. In these instances the new system may consist of an upgrade or extension of an existing system capability. Being required to use specific interface formats or specific existing hardware or software resources is an example of a constraint. The definition of constraints can take place in parallel with the definition of user needs.

In some new systems there may be a high degree of inheritance from previously successful systems. Such use of previously developed hardware or software or both can also be considered as design constraints at this point in the overall process.

While constraints represent implementation "hows," the major emphasis through the system requirements review still lies with the definition of "what" the system must do. The fact that how's are recognized early in the process does not mean that final system design is taking place at this point.

## WORK BREAKDOWN STRUCTURE

When it is determined that a sufficient set of materials representing user needs and constraints has been assembled, the systems engineer should next engage in an isolated, uninterrupted, and very challenging thought process.

The process involves developing a preliminary *work breakdown structure.* This is best initiated behind a closed door, following the procedures as given in Chapters 5 and 6.

## DEFINE DESIGN CRITERIA

Design criteria are developed by the systems engineer and finalized by the system design team under the leadership of the systems engineer. The design criteria consist of compiling *prioritized competing design characteristics,* a *design tradeoff methodology,* a *margin management program,* hardware and software analysis and design techniques to be used, and designing breadboard, brassboard, prototype, and early prototype strategies to minimize risk (these terms are defined in Chapter 12). Design criteria are documented in the systems engineering management plan.

## FORM THE SYSTEM DESIGN TEAM

With the systems engineering management plan and a preliminary work breakdown structure as guides, the systems engineer may now determine the representation by discipline required on the system design team. The design team consists of subsystem cognizant engineers, user representatives, and other required disciplines, such as logistics, test, production, and specialty engineering. There should be ten people or less on your design team. The design team is the focal point of the systems engineering process from this point on. The presence of actual users or capable user representatives on the system design team is of paramount importance. The team should meet once a week (see Chapter 7).

## SYSTEMS ENGINEERING MANAGEMENT PLAN

The system engineering management plan formally states how you intend to do business. The plan documents your design criteria, WBS, top-level schedules (including review strategies), approach to cost accounting, performance measurement, documentation standards, rationale for selecting an implementation paradigm (discussed in Chapter 2) and contingency (technical backup) strategies.

The systems engineering management plan is negotiated and agreed to by the system design team before any further work takes place. This may involve considerable skill on the part of the systems engineer, who by fortune possesses 51 percent of the vote. As with all documentation, feel free to reference other documents, such as the configuration management plan. Duplicating the same words in different documents is to be avoided as much as possible. Rewriting the same text in different documents dangerously increases the burden associated with change management (see Chapter 8).

The following four products of systems engineering can be developed in parallel.

## FUNCTIONAL REQUIREMENTS

The *functional requirements document* is the formal statement of "what" the system under development must do. Complete statements of "how" a system is to meet each functional requirement are contained in the *system specification document.*

An example of a functional requirement stating what must be accomplished is:

"The system shall accommodate sufficient storage for all personnel records over a period of five years following implementation."

An example of the corresponding specification stating how this requirement is to be met is:

"The computer memory shall include 180 megabytes of random access memory for the storage of personnel records."

The only "hows" permitted in the functional requirements document are those dictated by the implementation constraint requirements discussed earlier.

Every project *must* have a functional requirements document, even if it is only a few pages long for small tasks. If you are working on a project that does not have a functional requirements document approved by the user, stop work immediately and generate one. If you do not have the power to stop the work, look for another job immediately. A surprising number of systems are implemented without an adequate agreement or understanding of requirements.

Systems that are built without a concrete understanding of what they are supposed to do or that do not have constant feedback and approval from the ultimate user community are not likely to be successful nor can their implementation be successfully measured (see Chapter 10).

## CONFIGURATION MANAGEMENT PLAN

Configuration management is often considered a chore or inconvenience—something imposed on us by bureaucrats. The fact is that management of the configuration, through its various levels of evolving rigor, is a principal function of systems engineering management. Design maintenance, the orderly control of change, and a knowledge of deviation from the original functional requirements through waivers are all essential functions of the systems engineer.

Configuration management consists of four components—*identification, configuration control, auditing,* and *status accounting* (see Chapter 8).

## INTERFACE DOCUMENTATION FORM AND FORMAT

There should be one interface document or chapter in a single document for each internal and external system interface. The form and format for this documentation is defined at this point in the overall process (see Chapter 9).

## INTEGRATED LOGISTIC SUPPORT CONCEPTS

Components of the generic integrated logistics support process applicable to the specific project are determined at this time. Top level needs for supply support, test equipment, transportation and handling, technical data packages, support facilities, personnel and training, and the maintenance plan are defined (see Chapter 11).

## TRADE-OFF ANALYSIS

When functional requirements are completed, the system design team next turns to the question of "how" the system may be implemented. This is the first time "how" is diligently addressed. Alternative system designs are identified and studied in sufficient detail to demonstrate that system realization is technically feasible. Relative risk should be identified for each alternative. Tools typically employed in option analysis are queueing theory, simulation, costing, availability analysis, logistics support impacts, and so on (see Chapter 12). The system design team is the principal forum for identifying options and executing appropriate analyses. A result of option analysis is the recommendation that a specific option be implemented. The choice of the recommended option is made by rating each option with respect to the system prioritized competing design characteristics (see Chapter 6).

## THE SYSTEM REQUIREMENTS REVIEW

With functional requirements complete, a *configuration management plan* determined, *interface* documentation form and format defined, an understanding of *integrated logistics support* requirements, and a technically feasible design course established, the project is now ready for the first in-depth formal review with the customer. These items make up the central content of the *system requirements review.* A basic purpose of the system requirements review is to demonstrate that the sponsor's requirements are understood. This is accomplished by demonstrating a knowledge of *what* has to be done to meet the customer's needs and the identification of a feasible approach along with any attending risk.

If you pass the system requirements review, you earn the right to proceed with a *preliminary design*—that is, you are allowed to finalize a design option and develop *specifications*. Upon successful completion of the system requirements review, the functional requirements document is signed off and placed under configuration control. This level of design is referred to as the *functional baseline*. Procedures and plans for transition to the preliminary design review should also be covered at the system requirements review.

## SYSTEM SPECIFICATION

The system specification delineates "how" each function is to be realized. Functions are partitioned into hardware, software, and procedural functions. Detailed system requirements stating how each requirement for performance, physical characteristics, availability, environment, integrated logistic support, testing, design and construction are provided. Functional area, or subsystem, requirements for each of the above items are also provided in detail (see Chapter 10).

## SYSTEM TEST REQUIREMENTS

*System test requirements* are developed at this time for inclusion in the specification. System level test requirements state what tests are to be performed and must provide for the testing of all system functional requirements. The actual test design comes later in the documentation of *test plans and procedures*. A traceability matrix is included to demonstrate complete coverage of all functional requirements (see Chapter 14).

## PRELIMINARY INTEGRATED LOGISTICS SUPPORT PLAN

The *preliminary integrated logistics support* plan is developed at this time for inclusion in the specification. The level of detail is enhanced beyond the concept level to the complete specification level, but is still short of a detailed design (see Chapter 11).

## SOFTWARE REQUIREMENTS

The *software requirements document* is generated after the *system specification* development has proceeded to the point where system level software requirements have been clearly defined. The software requirements document partitions software into *computer program configuration items* and *computer program components* as a minimum and specifies their functions and interactions. In large or complex software designs, where implementation uncertainties may exist, computer program components may be further partitioned into *modules* and even *procedures*. Some software engineers prefer a software requirements document to be followed or supplemented by a software design document. Production of a software design document at this point is largely determined by the software system engineer's confidence that what is to be proposed at the upcoming preliminary design review is realistic, or by the software system engineer's sense that the course, to be followed in any subsequent detailed design, must be well articulated for his or her programmers. Software design tools and methodologies are also clearly specified at this time (see Chapter 13).

## INTERFACE REQUIREMENTS

The established form and format for interface requirements is now incorporated to produce requirements for all system and subsystem interfaces. These requirements include hardware, environmental, software, and procedural interface specifications. Hardware interfaces consist of mounts, cables, conduits, cages, hoses, and so on. Environmental interface requirements deal with electromagnetic radiation, thermal transfer, vibration, hostile environments, and so on. Software interfaces encompass information transfer at various levels driven by computer programs, and procedural interfaces refer to requirements for human interactions (see Chapter 9).

## MAINTAIN FUNCTIONAL BASELINE

Throughout the period from the system requirements review to the preliminary design review, the functional baseline documentation, which consists of the functional requirements and the configuration management plan, are subject to configuration management. This function is carried out by the *configuration management board* and is supported by the systems engineer and the systems design team (see Chapter 8).

## THE PRELIMINARY DESIGN REVIEW

With complete system-level specifications and specifications for software, system-level, and subsystem interfaces documented, preparation begins for a formal review of the preliminary design. The purpose of the preliminary design review is to demonstrate the feasibility of the chosen design. At the preliminary design review, system specifications, test strategies, software requirements and specifications, the preliminary integrated logistics support design, and appropriate technical management strategies for the detailed design phase to follow are presented. This level of design is called the *allocated baseline.*

Passage of the preliminary design review earns the right to proceed with a detailed design—that is, the team is then allowed to begin actual fabrication.

## SUBSYSTEM DETAILED DESIGN

Hardware and software partitioning has been completely defined by the system specification and software requirements documents in terms of functions to be performed and top-level interface responsibilities. Hardware subsystems are referred to as *hardware configuration items.* Software programs at this level of decomposition are referred to as computer program configuration items. Detailed designs will result in the production of subsystem specifications for hardware and the completion of the design section of the *unit folder* or its equivalent for software. These documents include derived unit test plans and procedures traceable to unit requirements. There are no consistent rules for the level of detail presented in these designs. The level of detail and the effort required to produce the acceptable level of detail is determined by the level of confidence in the ability to implement the design. If the item is "off the shelf" or an exact duplication of an existing item, the design effort and supporting documentation may be minimal. If the item in question has never been built before, tools such as breadboarding, brassboarding, simulation, and prototyping may be employed to gain the necessary confidence, prior to the critical design review, that implementation will be successful.

## INTEGRATED LOGISTICS SUPPORT DETAILED DESIGN

The integrated logistics support detailed design for the product is accomplished during this phase. The design is documented in the integrated logistics support plan (see Chapter 11).

## INTEGRATION, SYSTEM, AND ACCEPTANCE TEST REQUIREMENTS

Test requirements documents specify all tests to be carried out. The set of tests for integration testing is derived from the *hardware configuration item* and *computer program configuration item* interface requirements and specifications. The set of tests for system level testing is derived from the system functional requirements document.

System tests are tests that the implementation organization conducts to validate and verify system performance. Traceability matrices assure that at least one test for each specification and functional requirement is included.

Acceptance tests are tests that the customer exercises as a criteria for system acceptance. In simpler systems, the acceptance test may consist of a repeat or observation of the system level tests. In more complex systems, this may not be practical and statistical tests may be employed. In any case it is important that acceptance test content be negotiated early enough so that surprises don't arise just before delivery—that is, agreed-to acceptance tests should always map into or onto the implementor's system tests (see Chapter 14).

## PRELIMINARY TEST PLANS AND PROCEDURES

Test plans state how the test requirements are to be met. They state whether tests will be performed by inspection, analysis, or actual performance measurements. They also include the sequence of testing and the types of hardware, software, and personnel resources required.

Test procedures provide a detailed account of each step to be carried out for each test—when the tests are to be performed, by whom, the precise equipment and personnel support required, predictions of results, and procedures to follow in the event that anticipated results do not occur. Preliminary test plans and procedures provide the form and format for the final test plans and procedures to be developed after the critical design review, as well as providing as much detail as may be possible while detailed design is in progress. Test plans and procedures cannot be finalized until the detailed design of all hardware configuration items and computer program configuration items is complete. The preliminary test plan and procedures, however, must provide sufficient confidence to reviewers at the critical design review that all testing will be sufficient to validate and verify system performance (see Chapter 14).

## INTERFACE DETAILED DESIGN

Detailed design of hardware configuration item and computer program configuration item and system interfaces takes place in concert with hardware and software detailed design. System and subsystem interface boundaries constitute an exclusive province of responsibility for the systems engineer (see Chapter 9).

## MAINTAIN ALLOCATED BASELINE

At the preliminary design review, all system functions were partitioned and allocated. The principal function of configuration management during this phase is to maintain the allocated baseline design. Configuration management becomes increasingly stringent as defined baselines proceed. A balance is struck during detailed design. Major deviations, surprises at this point, are rigorously controlled. At the same time, the configuration management process cannot be so rigorous as to impede freedom in details of design choice, particularly when breadboarding, simulation, and prototyping may be taking place. It is a challenging time for the systems engineer and configuration management support personnel who must guard the integrity of the allocated baseline, yet allow for creativity and unrestrained progression of detailed design.

## THE CRITICAL DESIGN REVIEW

The critical design review is appropriately named. The purpose of the critical design review is to demonstrate that the design is mature and ready to implement. Schematics, engineering drawings, data flow diagrams, structured designs, integrated logistics support designs, testing designs, and so on, are sufficiently complete so that implementation risk is well understood and acceptable.

Confidence of both the implementors and the review board must be high to pass the critical design reviews. Passage of the critical design review earns the right to actually build the system—that is, cut metal and begin formal coding.

In complex systems, there may be more than one critical design review. The systems engineer may call for one such review for hardware and another for software, or may call for multiple reviews at this level for subsystems. A critical design review can last for hours or for days. Again, this depends totally on one's level of confidence in design detail. As a rule, code production should not begin until the Critical Design Review is complete. Exceptions to this rule may be tolerated if some code production is required for early prototyping, in order to clarify requirements or to justify the realism of specific algorithms, and so on. However, this type of code production must not get out of hand, even if some of the work may be inherited into the final product. The perspective must always be maintained that prototype code supports the design process as needed and that implementation in earnest does not begin until successful completion of the critical design review. One of the toughest jobs facing the systems engineer is to keep design before implementation.

Successful completion of the critical design review establishes the *design baseline* under configuration management.

## CODE AND UNIT TESTING

One of the virtues of using tools such as structured analysis, structured design, and unit folders (or their functional equivalents) from the systems engineer's viewpoint is that these tools can result in the production of documentation as the software implementation progresses. Coding is the most fun that programmers have. If programmers are allowed to complete coding before any documentation is produced, the probability of ever obtaining adequate documentation is greatly decreased.

Under these conditions, I have actually seen programmers threaten to quit if they had to "stop work" and produce documentation after the fact. To the systems engineer, software includes code *and* documentation.

The selection of software tools can be very important. When properly used, they can actually help programmers. They also provide a structured visibility for the systems engineer and the software systems engineer. The selection should be jointly agreed to by the system design team, but should meet the minimum criteria of the simultaneous production of code and documentation. Code walkthroughs and peer reviews are the major systems engineering tools for maintaining visibility at this time (see Chapter 13).

The importance of thorough unit testing cannot be emphasized enough. Three basic rules of testing are test early, test early, and test early (see Chapter 14).

## PROTOTYPE BUILDING AND TESTING

On the hardware side, the term prototype refers to units that are very close to production units and is often used to refer to serial number 1 itself. If it is not the actual unit planned for production, it is close enough to serve as a final test unit prior to larger-scale production commitment. Hardware people are more used to completing designs before they build. Again, the need for thorough hardware unit testing cannot be overstressed.

## FINAL TEST PLANS AND PROCEDURES

With the detailed design for all hardware configuration items and computer program configuration items complete, test plans and procedures at the subsystem, integration, system, and acceptance test levels can be finalized in parallel with implementation (see Chapter 14).

## SYSTEM INTEGRATION AND TEST

System integration involves the methodic consolidation of subsystems into a system whole. The sequence must be carefully planned to support the testing function in a complete and efficient manner.

The effort concentrates on the behavior of interfaces. When failures occur, it is of the utmost importance to be able to isolate the culprit. Thus, subsystems must be designed not only to perform their allocated functions, but to efficiently support the integration test process. Designing for testing (that is, the inclusion of efficient test points and consideration for the information they will provide) is an important part of system design. It is a fact of life that it is never really clear how well things are going until system integration and test is launched (see Chapter 14).

## SYSTEM TESTING

With the successful completion of integration testing, which concentrates on the performance of system and subsystem interfaces, the complete end-to-end system is now ready for system testing. System-level tests are designed to validate the system's ability to meet system function requirements (see Chapter 14).

## PRODUCTION TEST AND PRODUCTION

There are two major rationales for production testing. The first is to assure that the initial production units behave the same as the prototype units that were subjected to prior integration and system-level testing. These may or may not be identical units. In either case, the production environment is different and early validation is required.

The second reason has to do with ongoing quality control throughout the production life cycle. Historically, the degree to which an organization invests in production quality control through statistical testing has depended very much on the product and marketing philosophy for the product. More recent tenets of *total quality management,* or TQM, call for the use of statistical process control methods to continually monitor and improve the production process, greatly reducing the need for the more traditional approach of inspection after the fact (see Chapter 14).

## SYSTEM ACCEPTANCE TESTING

Acceptance tests are tests conducted by the customer, by the vendor with the customer present, or a combination of the two. Successful completion of the tests is designed to define a specific point in time at which the product baseline is officially transferred to the customer. Completion of successful acceptance testing generally defines the point in time when maintenance contracts come into effect.

Acceptance tests are often conducted at the installation site and are coordinated with system phase-in activities. When the installation site is different from the production site, pre-shipment and post-shipment tests are often performed by the vendor prior to customer acceptance testing (see Chapter 14).

## FINAL DOCUMENTATION

All required system documentation is produced in accordance with the documentation tree and is updated as required throughout the design, implementation, and test processes. Additional need for changes often occur during system, production testing, and acceptance testing. A final review and updating of all system documentation should take place in parallel with acceptance testing.

After the critical design review, configuration management becomes increasingly strict. The *change control board* is in full swing, and any deviations from the design presented at the critical design review are considered serious. Requested changes are thoroughly scrutinized for their programmatic and technical impacts. The systems engineer and the system design team play a substantial role in supporting the control board during this time.

# 4

# Defining User Needs and Constraints

## WHO ARE THE USERS?

Some years ago an information system was installed in a hospital in England. The system concept was to gather and provide data at the nursing stations in each wing via CRT terminals and hard copy devices. In the process of placing orders for supplies, clinical tests, and other services, data was to be integrated into a central computer for purposes of central billing, monitoring inventory, providing updated patient records, alerting admissions and other services of impending events, scheduling surgical suites, and compiling consumer demographics and other statistical reports. The CRT displays were menu driven.

One of the problems encountered was that space at the nursing stations was extremely limited. The engineers decided to put the terminals on carts with casters so that they could easily be moved about when access to specific areas was required.

After two weeks of training following "installation," the nurses had managed to push every terminal on each wing of each floor down to the end of the corridors and leave them there. What happened?

Throughout the entire user need and requirements generation phase of the project, not one nurse (user) was involved in the system design. The menu-driven system required seven steps through medical ordering submenus just to order aspirin. While menu-driven systems have many excellent applications, this was not one of them. The engineers never seriously questioned the applicability of this technology to the nursing setting. Routine nursing clerical tasks were suddenly perceived as taking an excessive amount of time, with no apparent benefits. Benefits of the system design were oriented toward more efficient management. The motivation of most nurses is oriented toward efficient direct patient care—not necessarily toward management. Fortunately, the previous vacuum-driven tube system through which documents were transported in cartridges was left in place as an emergency backup system. In short order, "the user" decided it would once again be the primary system.

It is, of course, incredibly easy for us to be pontifical about what happened in this example. User needs and requirements cannot be developed in a vacuum. Unfortunately, what happened in this instance is not an isolated event. It is, in fact, a very common event.

The term "user" actually refers to everybody in the customer organization. This includes top management because they have to perceive a benefit and foot the bill. It includes middle management because they have specific department or division-oriented goals. It includes line management and supervisors because they have well-defined group goals. And it includes the people who actually touch the system on a day-to-day basis because they typically have consumer interface–oriented goals. These goals can be, and are, extremely diverse. If a system design fails to meet all of them, it is clearly deficient. If the day-to-day user rejects it, it is a total failure.

Evidently, the most prevalent reason for system implementation failure is that the system as delivered simply does not do what the customer perceived was needed. There can be many reasons for this—changes in requirements, drifting off course during design, misinterpretation of what users say—but the most unforgivable reason is that *all* of the "users," including the day-to-day users, were not consulted in the beginning and kept abreast of changes throughout the implementation process.

## WHAT TO DO?

User needs are not functional requirements. Developing functional requirements, a highly structured statement of exactly what the implementor proposes to do, comes later. Gathering user needs is a less structured, liberal exercise. User needs are compiled from data gathered from existing documentation and from personal interviews. It is basically a complete wish list derived from the stated needs of the entire spectrum of the target user community.

Stategies for determining user needs are greatly influenced by the size of the user community. For systems that involve high levels of production, such as in the automobile and consumer electronics industries, it is not possible to interact with all of the users. User needs must therefore be gathered by statistical means through market research. One of the best ways to do this is to conduct interviews on existing products in an effort to determine likes, dislikes, and absent desirable features. It is also useful to field test prototype concepts at trade shows and conventions where the customer can touch and feel.

Systems engineers are rarely experts in conducting market research, and typically need to rely heavily on these resources within their organization. Unfortunately, there can be considerable pressure brought upon the systems engineer by other organizational interests, such as costs of development and production. It is not uncommon for management, engineering, and production personnel to override the systems engineer with development decisions that are inconsistent with the findings of market research. If you have a choice, go with what the customer wants. If you don't have a choice, your organization has a quality problem[2].

When the customer community is relatively contained, the process may be simpler but no less difficult. Examples of such systems are a spacecraft, a law enforcement communications system, or a government information or hardware system. In systems of this nature, the first step is to read all available documentation, including customer-generated requirements documents, specifications, internal memos, meeting minutes, engineering studies, or any other formal or informal material that refers in any way to the history of the perceived need. Usually by the time a user organization and a vendor organization first come together, a number of documents have been generated by the user in the normal course of maturing in their own thinking to that point.

Commonly, the first view of the customer's need comes in the form of a *request for proposal* (RFP). The RFP may or may not ask for development of requirements. If it does, you're determining user needs up front. Often, the RFP attempts to specify requirements. This can be an ambitious task for private sector and government organizations that do not routinely procure the type of system in question. For example, a stock exchange that procures a modern transaction handling or data display system is likely not to have initiated such a procurement in a number of years, if ever. They cannot be expected to have top-notch technical RFP generators on staff and would be well advised to use a consultant.

Alternatively, branches of the military exhibit considerable experience in generating sophisticated RFPs due to their frequency of procurement. The experience of the procurer is usually evidenced by the quality and completeness of their RFP. If the requirements seem complete and are measurable and the level of experience of the requestor satisfies you, the need for a *user needs gathering phase* may be greatly reduced or even eliminated.

The technical portion of a well-written RFP should contain a background statement, a succinct statement of the problem, top-level functional flow diagrams, system constraints, measurable requirements in terms of what has to be done, and a clear statement of work. If you are wary with regard to the quality of the RFP, there is nothing wrong with proposing that a thorough review of requirements through a user needs assessment be conducted as a first task.

The second step in gathering user needs is to conduct interviews. Interview design is something that engineers are generally not experienced in. There is an entire discipline devoted to it. For example, it is best to ask general open-ended questions up front and gradually move toward specifics. Like all of us, respondents don't like to contradict themselves. Establishing concrete positions too early can influence degrees of freedom in later responses. Different statements of the same question are often inserted out of context in an effort to corroborate earlier responses. Open-ended sessions, where questions tend to migrate in their manner of presentation or even change as the interviewer gets "onto something," must be conducted with great care or else consistency may be sacrificed. There is a definite art to the business. The best thing to do is to get help with your questionnaire design from elsewhere in your organization or from an outside professional.

When you think you have a good questionnaire, review it with your customer. You will most likely have one questionnaire for top management and other questionnaires for middle management, line management, and the hands-on system users.

Next, field test the questionnaires with a representative sample of respondents. Limited preliminary field testing of questionnaires invariably uncovers deficiencies, such as omission of specific questions related to unforeseen user perspectives, possible misinterpretation of questions, or the unintentional skewing of answers by the set of immediately preceding questions. These are very common deficiencies in questionnaires.

The construction and execution of questionnaires is a typical area in which systems engineers may want to seriously consider the support of outside experts.

## FACTORS THAT MAKE IT DIFFICULT

Barry Boehm offers an informative list of factors that contribute to the difficulty of gathering user needs and formulating requirements [1]. The list is given in Table 4-1.

While one needs to be aware of all these impediments, item 5 is of particular interest. It is common for people to verbalize their problems in terms of perceived solutions. It is very important to be sensitive to this distinction early in the process.

For example, the following statements are solution statements, not problem statements:

**TABLE 4-1 Factors that Make Determination of User Needs and Requirements Difficult**

1. It's not always clear who your customer or user is.
2. Your customer cannot always state his requirements clearly or state his requirements completely.
3. What your customers say they want may not be what they need.
4. You may begin to give them what you think they need, instead of what they want.
5. Their concept of the solution may not solve their problem.
6. The customer or user will have implicit expectations that may be unreasonable—and unknown to you.
7. Not all of your customers or users talk to one another, so you must talk to all of them.
8. Your customer or user may change.
9. If the customers or users are not part of the planning and requirement analysis process, they may be reluctant to accept the product.

"We need to put in an East-West teleconferencing network."
"Management needs the ability to conduct quick-look 'what if' studies related to resource allocations."
"We need a computerized outpatient management system."
"We need a dispatching system for our trucks."

Additional probing in response to such perceived statements of need is required to extract more generic statements of what the problems really are. Truer problem statements associated with the above examples might be:

"Our East-West offices are not communicating in a timely manner."
"Our resources are becoming oversubscribed. We need to increase our efficiency with regard to resource allocation."
"We are not able to track our outpatients with sufficient precision to provide needed checkups in a timely manner."
"We need to improve the efficiency of our use of vehicles in the field."

Authentic needs statements are identifiable because they logically give rise to further questions that result in needs that are measurable. When beginning with a true needs statement, one can sense this happening. Given the East-West communication problem, for example, one may now begin to pinpoint exactly what kind of information is not being exchanged in a timely manner, what its precise content is, and what statistical response times are desired.

Avoiding "jumping to solutions" is not always easy. Ideas (solutions) related to our individual technological expertise naturally flourish in our minds. There are times when "solutions" seem so clear that it requires great discipline to slow down and concentrate on what the problem really is. At this stage of the game, we are not a bag of solutions looking for problems. We are taking the first step toward problem definition—we are listeners who guide the customer toward generic problem understanding.

To lessen these complications further, the implementation team does well to embrace the customer's cause by making an honest effort to visualize themselves in the user's occupational specialty. People will seldom support a design they are not involved in. To get them involved, we must get involved. We must learn their terminology and seriously prepare to be conversant so that our questions and insights are meaningful. We must concentrate on *what* they need and postpone our urges to implant our pet technologies early in the process. It is important, of course, to guide a respondent should he or she ask for features that are technologically infeasible or involve excessive cost. But our main duty at this point is to *listen*. Most user communities have one important quality that implementors don't have—they know their job better.

## SUMMARY

The following guidelines are offered in support of the gathering of user needs:

1. Read everything you can find in preparation. Become as conversant as you can with the complete user environment. Ask questions. Change your shoes. Get smart.
2. Construct interviews for top management, middle management, supervisors, and hands-on users. If you have any doubt about your ability to construct a questionnaire, get help.
3. Review your questionnaire with the sponsor.
4. For each basic function in the user environment, try to document top-level functional statements and needs for performance by measurable criteria, system design constraints, availabilities, logistics support, and information transfer in terms of data types, volumes, formats, response times, and tolerable error rates.
5. Field test your questionnaire. Fix it and try it again. Don't use it until it works.
6. Don't be omnipotent. In his or her working environment, the customer is smarter than you are.
7. Care about the people whose problem you are trying to understand. The more you care, the more they will like you. The more they like you, the better they will communicate with you. The better they communicate with

you, the better chance of success you have. In this stage of events, you're the one that needs help.
8. Don't tell stories designed to impress the customer with your knowledge. *Listen, listen, listen.*

## CONSTRAINTS

System designs are often constrained by existing systems that they need to interface to. Rarely is there a brand-new system built from scratch that is not impacted by some other system. The discernment of constraints should be an integral part of the gathering of user needs and requirements formulation. Constraints should be clearly documented in the functional requirements document. Constraints are the only "hows" permitted at the functional requirements level.

One familiar constraint is the need to use existing equipment, commonly at specific system interfaces. For example, consider a system to transmit a newspaper published in New York on a daily basis for publication in Chicago. Typical functional requirements may include the need to send copy after 2 A.M. upon completion in New York and to accomplish all copy transfer by 4 A.M. in time to set up presses in Chicago. (This requirement sets data rates.) Further requirements may include specification of error rates, system availability, data types, personnel and training, and so forth.

Note that these requirements state "what" must be done. The determination of "how" (satellite or land communications, use of aircraft or carrier pigeons, etc.) is yet to be determined during options analysis and preliminary design. However, a typical constraint in this example may be that the data interface in Chicago must use a particular computer model and that the final product must be in a particular format on a 1600-BPI magnetic tape.

There are always cost and schedule constraints on any project. In some cases, however, they are absolute drivers. Examples are design-to-cost projects and space-borne systems that may need to be completed in time for specific launch windows of opportunity.

Severe constraints may also be imposed on commodities such as power, weight, size, and levels of performance, or by design inheritance. While issues of this type may blur into classification as functional requirements, an early understanding of their potential as design drivers should be sought—particularly when they clearly affect the "Hows" of implementation.

While user needs and functional requirements deal with what has to be done and not how it is to be implemented, it is quite appropriate to include a section in the functional requirements document that deals with constraints. This is the only section at the user need level or at the functional requirements level in which it is permissible to talk about how. What is being stated by such an inclusion is, "When you do this part of what we want you to do, you must do it this way."

It is up to the systems engineer through interaction with respondents to determine if the constraint under discussion is truly to be a design limitation. If it is, identify and understand it as a valid part of the user needs exercise and describe it in any level of detail that is required.

**References**

1. Software requirements analysis and design. 1985. Data Processing Management Association Seminar, San Jose, Calif., July 1985.
2. Reilly, Norman B. 1993. *Quality: What Makes it Happen?* New York: Van Nostrand Reinhold.

# 5

# The Work Breakdown Structure

Having defined and published a *user needs document,* the systems engineering process next comes to the point where a first cut at determining what has to be done is made. The tool to be used is the *work breakdown structure* (WBS). The WBS is often constructed by project management. The systems engineer should strive to be a major player in this effort and may do so by volunteering to make the first cut.

The WBS will wield a prodigious effect on system architecture, subsystem definition, system interface definitions, how you organize, what gets scheduled, what costs get tracked, the ability to execute performance measurement, and the definition of required disciplines to succeed.

If someone else has already built a WBS, *you should still go through the following structured thought process!*

## WHY A STRUCTURED PROCESS IS REQUIRED

At the inception of any system, there is a famous point in time. It is the time when a project manager, a project engineer, a user generating an RFP, a systems engineer—whoever—takes paper and pencil into hand and begins to bring structure to his or her thoughts as to what has to be done to attain a total system. There is no widely accepted standard on how to do this. In fact, it is rare for any two people faced with this problem to proceed in the same fashion.

Some start out by trying to draw block diagrams. Others begin by making schedules or drawing organization charts or attempting to construct functional flow diagrams. Approaches such as these can be dangerous because they are typically oriented solely toward the product. In doing this, the analyst automatically fails to engage in a thought process that gives appropriate attention to issues such as testing, integrated logistics support, management, and systems engineering in a consistent top-down integrated fashion. An early concentration on the product alone leads to uneven emphasis with regard to the total system. It does not inherently provide a generic methodology to help one think of everything that needs to be considered and to decompose the work at consistent levels of description. Omissions, if they are caught, become ungainly appendages.

Further, these approaches rely heavily on one's experience. Experience, of course, is fundamental, but it can represent a serious drawback. In the absence of a methodology for a structured decomposition of work to be done at consistent levels, there is an overwhelming tendency to emphasize those technical areas with which one is most familiar and to slight or even forget those areas with which one has less familiarity. This often results in excessive detail in one area and shallow treatment or omission in another area.

There is also one other extraordinary error committed in attempting these approaches—that is, neglecting the fact that the wheel has already been conceived and created. The system structuring process has been attempted before—so many times, in fact, that over 20 years ago substantial effort was devoted to developing a comprehensive guideline for the organized decomposition of work. It is called MIL-STD-881, "Work Breakdown Structures for Defense Material Items," first produced in November 1968. Defense contractors are, of course, totally familiar with the latest versions of 881; the private sector is generally less aware of its benefits as a guide for completeness. The process is equally effective for nondefense systems.

The decomposition process presented in this chapter is based on the 881 approach. The work decomposition process is functionally the same as the decomposition processes encountered in decomposing a software system into computer software configuration items, programs, and so on, the decomposition process carried out in knowledge engineering, and the decomposition process of structuring the organization of a hierarchical data base.

It is a top-down process in which each successive level of detail consists of naming disjoint sets whose sum maps into and onto the entire problem space at each step. Each level of detail is directly traceable to an entity at a higher level of decomposition. Everything has a specific, well-defined place. It can be a difficult process requiring many iterations, but when done correctly the structure will exhibit a peculiar beauty. That beauty is most severely tested when an unforeseen addition is required long after the structure is in use. The robustness is evident when the new addition does not perturb the structure, but comes immediately to a resting place that is logical and consistent—a resting place that is unmistakably identified by the structure itself.

It is always best that the first cut at a decomposition process be carried out by a single person, in isolation, with sufficient time and a minimum of pressure. It is expedient for the systems engineer to volunteer to make this first cut. There will be time for review and critique by the system design team and by project management. Certainly, as systems engineer, you will want to have as much influence as possible on the architecture and content of the WBS. Further, when confronted by a WBS that is inherently incorrect, the systems engineer will want to make appropriate recommendations for change based on logical and structured reasoning.

The WBS presented in this chapter is a generic one. The term "generic" suggests that the structure applies to the development of the entire class of "systems"—that is, any system. The basic structured thought process involves addressing each item in the generic WBS at successive levels of detail. The generic guide is intended to generate complete horizontal consideration of the entire system at each level of detail *before proceeding to the next level of detail.* Review and iterations between levels are, of course, likely to occur. At any given point in time, however, the principle concentration of effort is devoted to completing a specific level.

It is recognized that there are differences between work breakdown structures built for development, for production, and for operational environments. The material presented here is primarily devoted to systems engineering during system development phases. It is understood, however, that a system is designed for operations and that operational considerations drive requirements for the mission product and its logistics support.

## THE WORK BREAKDOWN STRUCTURE FOR SYSTEM DEVELOPMENT

Table 5-1 presents the top two levels of the generic WBS. The first level is the system level, which names the system. The second level lists the five components *always* present in the development of any system:

1. The mission product is simply the complete entity that is to be created. The mission product consists of the hardware, software, auxiliary equipment, and integration equipment items that make up the system product itself.
2. Integrated logistics support addresses how to support the development of the mission product and how to support the mission product in the operational environment.

**TABLE 5-1 The Generic WBS**

| Level 1 | Level 2 |
|---|---|
| System | |
| | Mission product |
| | Integrated logistics support |
| | System testing |
| | Project management |
| | Systems engineering |

3. System testing refers to the functions of validation and verification of system performance.
4. Project management covers programmatic issues.
5. Systems engineering is responsible for the technical success of the fielded mission product and its support mechanisms.

## The Mission Product

Table 5-2 presents examples of typical mission product items at level 3. At this level of the thought process, the analyst concentrates on basic functions needed to meet the stated user needs. The level of consideration is important. We are at level three of a top-down process. It is a level at which basic

**TABLE 5-2 The Generic WBS—Mission Product**

| Level 2 | Level 3 |
|---|---|
| Mission product | |
| | (examples) |
| | Sensors |
| | Communications |
| | Navigation |
| | Computers |
| | Computer programs |
| | Workstations |
| | — |
| | — |
| | Propulsion |
| | Fire Control |
| | Armament |
| | Assembly |
| | Auxiliary equipment |

*Note:* You always have the last two items.

technologies and disciplines are considered. If we perceive that light detectors are required, we do not put down light detectors—we put down sensors. If we perceive radios are required, we do not put down radios—we put down communications. Similarly, we put down navigation, if required, at this level, not inertial navigation platform. Nor do we put down items such as diesel engines or ram-jets, but instead we put down propulsion. In short, at this level we are still addressing fundamental functionality. The "how" comes later.

The last two items listed in Table 5-2 warrant particular consideration.

Assembly equipment is that equipment employed to integrate, or connect together, the other level 3 functional items. This includes such items as cables, conduits, racks, mounts, and so on. Every system has assembly equipment. Assembly equipment will *always* be an item at level 3 of your WBS. Engineers that design and build functional level 3 items are typically not interested in how they get connected with other level 3 items to form a system. The systems engineer must be.

Auxiliary equipment is that equipment to be used by more than one level 3 functional item or by personnel in the operational environment. This includes such items needed for environmental control, production and delivery of electricity, common storage space, common work space, security, shelter(s), and other commonly used facilities. It is rare that a system does not have some type of auxiliary equipment. If none occurs to you at this level of the thought process, you will still do well to include it for completeness. It is easier to remove items from a structure than to insert them after the structure is formulated.

The structuring of a WBS is commonly prone to the making of two types of errors. The first is compromising the top-down nature of the effort by mixing different levels of detail at any given level. The second consists of errors of omission. The generic WBS guide is designed to minimize the probability of committing either type of error.

Development of requirements, design, fabrication, and test of level 3 items will eventually become the responsibility of cognizant engineers (COGEs), or subsystem engineers.

For each mission product level 3 item, level 4 COGEs should provide:

- System functional requirements support;
- System specifications support;
- Subsystem requirements;
- Subsystem specifications;
- Software requirements documentation;
- Software design and documentation;
- Hardware design and documentation;
- Fabrication (build/code);

- Unit test (hardware and software);
- System-level testing support; and
- Design team support.

Generation of specific sections of the system functional requirements and system specifications documents will be assigned to appropriate COGEs.

Subsystem requirements and specifications, which may be required on larger projects, are generated by COGEs in response to partitioning directives provided in system level documentation. Software requirements documentation is also generated in response to higher-level partitioning of functions, either at the system or subsystem levels. Design documentation responds to requirements or to specifications, should the latter level of detail be called for. Hardware and software fabrication level 4 line items are followed by unit test line items. Level 3 units are units that will be submitted to system integration and system integration testing. COGEs also support system integration, system testing, and system acceptance testing, and devote considerable time to system design team support throughout the development cycle.

## Integrated Logistics Support

Table 5-3 lists level 3 ILS items that must be considered in any development WBS. At this point of WBS structuring, there is no thought required in listing level 3 items under ILS in your WBS. Simply list them as shown in the table at level 3. If later analysis results in one or more of these items not being required or their accommodation elsewhere in the WBS, they can be eliminated at that time.

Supply support refers to the logistics function of providing equipment and parts directly related to the development and operations of the mission product. Test equipment is equipment needed to support all unit, integration, system, and acceptance testing. Transportation and handling covers all design

**TABLE 5-3 The Generic WBS—Integrated Logistics Support**

| Level 2 | Level 3 |
|---|---|
| Integrated logistics support | |
| | Supply support |
| | Test equipment |
| | Transportation and handling |
| | Technical data packages |
| | Facilities |
| | Personnel and training |
| | Maintenance plan |

issues related to the need to transfer the mission product from the delivery site to the operations site. Facilities are such items as structures and power plants required for the support of development activities. Required operational facilities are typically covered under mission product items. Personnel and training refers to the identification of specific personnel needs for development and for operations, personnel acquisition, and the identification and development of instructional materials and media. The maintenance plan provides the procedures and policies for logistics support and execution of all maintenance during development and operations.

Clearly, there are two distinct logistics support plans that must be generated—one to support development activities and another to support field operations. The former is generated to support the design and implementation process, and the latter is a part of the design process for support of the mission product itself when deployed for use.

Components of ILS for system development and for system operations are further discussed in Chapter 11.

### System Testing

Table 5-4 presents six major WBS items that constitute minimum system testing functions to be carried out. These are based on system-level integration testing, system testing, and acceptance testing. These items, discussed in more detail in Chapter 14, will always be included in the WBS.

### Project Management

The lines between systems engineering, systems engineering management, and project management are subtle, and their delineations are often a matter of style and experience or organizational policy. The style is often set by

**TABLE 5-4 The Generic WBS—System Testing**

| Level 2 | Level 3 |
|---|---|
| System testing | |
| | Integration test plans and procedures |
| | System test plans and procedures |
| | Acceptance test plans and procedures |
| | Integration testing |
| | System testing |
| | Acceptance testing |

project management. The items listed in Table 5-5 are generally under the province of project management. It is of extreme importance that the specific roles of project managers, deputy project managers, project systems engineers, software managers, and others be well defined and synergistic with regard to the project management WBS items.

In the generic sense, systems engineering supports project management in all technical aspects and in programmatic areas, such as costing and scheduling of technical activities.

## Systems Engineering

Systems engineering is the focal point of technical responsibility for the definition of user needs, requirements, design, implementation, test support, and successful delivery of a system to the operational environment. The generic level 3 line items that need to be addressed by systems engineering are listed in Table 5-6 and discussed in the following paragraphs.

### *Project Management Support*

The systems engineer routinely provides support to project management in the areas listed in Table 5-5. Good project managers will clearly define the roles of all personnel on a project in writing. When this ideal is not realized, it is wise for the systems engineers to develop their own statement of roles and objectives for project management review. The statement should include a paragraph covering each of the items listed in Table 5-5 and 5-6. For each item, specify the extent you intend to provide support and/or leadership. For example, the systems engineer will provide support in constructing and presenting project reviews, but will assume leadership in the organization and

**TABLE 5-5 The Generic WBS—Project Management**

| Level 2 | Level 3 |
|---|---|
| Project management | |
| | Objectives |
| | Policies |
| | Deliverables |
| | Reports/reviews |
| | Documentation tree |
| | Organization |
| | WBS |
| | Costing |
| | Scheduling |
| | Performance measurement |

**TABLE 5-6 The Generic WBS—Systems Engineering**

| Level 2 | Level 3 |
|---|---|
| Systems engineering | |
| | Project management support |
| | System design team |
| | User needs |
| | Systems engineering management plan |
| | System functional requirements |
| | Trade studies |
| | System design requirements |
| | Software management plan |
| | System software requirements |
| | Configuration management plan |
| | Interface documentation |
| | Test requirements and evaluation |
| | Specialty/concurrent engineering |

running of the system design team. Similarly, the construction of the software management plan may be assigned to a software systems engineer. In large organizations, this is likely to be the case. If so, the systems engineer should have review and approval authority.

The trade-offs between providing support and assuming the leadership in these items is largely a function of organizational size, policy, and the authoritative strength of project management on a given project. In the absence of lucid role definitions, it is advisable to negotiate clearly defined roles in each indicated area and ask that a statement of the responsibilities of the systems engineer be distributed to all project members. This avoids the danger of later confusion and even conflict, with regard to both the allocation of responsibilities and the authority to meet those responsibilities.

Major systems engineering roles and methodologies are also defined in the systems engineering management plan, which is discussed in Chapter 6. In any case, it is of extreme importance that the position and functions of the systems engineer be well defined on a project-by-project basis and that all parties concerned understand the extent of authority associated with each of those roles. It is a fact of life that you must have the authority to execute your responsibility.

### *System Design Team*

The system design team (SDT) is the forum through which systems engineers accomplish the great majority of their goals. It is the responsibility of the systems engineer to form and lead the SDT. Chapter 7 is devoted to the organization and management of this most important focal point of systems engineering activity.

***User Needs Documentation***

Ideally, the systems engineer will be given responsibility for compilation and production of the user needs document. It is not uncommon, however, that some kind of user needs assessment has taken place prior to the actual formation of a project or the specific assignment of a systems engineer. If this is the case, one of the first tasks of the system design team should be the review and formalization of this material into an updated user needs version as required. The systems engineer, along with project management and, of course, the user or a user representative, should then have signature authority over the new version.

***Systems Engineering Management Plan***

The systems engineering management plan (SEMP) defines how the function of systems engineering will be organizationally structured and how systems engineering will exercise technical control over the complete engineering process. The format and content of the SEMP is presented in Chapter 6.

***System Functional Requirements***

The systems engineer has prime responsibility for the creation of system functional requirements document (see Chapter 10). Project management, the user, and the systems engineer must sign off on this document as a minimum. As systems engineer, if you don't have the responsibility to produce this document and to control its content and the delegation of its creation, then someone else is the real "systems engineer."

***Trade Studies***

The systems engineer has complete responsibility for the identification and execution of all design trade-off studies. The methodology for trade studies is defined in the SEMP. Trade studies are executed by the system design team under the leadership of the systems engineer. Expertise external to the permanent system design team membership may be called upon as required. Methodologies for the study of options are discussed in Chapter 12.

***System Design Requirements***

The system specifications document establishes "how" each requirement in the functional requirements document is to be satisfied. The organization and construction of this document is the responsibility of the systems engineer (see Chapter 10). This responsibility includes signature authority.

***Software Management Plan***

This document is typically produced by the systems software manager or systems software engineer. The document is approved by both project management and the systems engineer. The systems software manager must work for and report to the systems engineer through the SDT. If this is not the case, a serious potential for conflict exists (see organizational structures in Chapter 15).

***System Software Requirements***

This document is produced under the direction of the system software manager. System software requirements are generated in response to the system design requirements, or specifications. The document must be approved by the systems engineer and may be approved by project management.

***Configuration Management Plan***

In large organizations, generation of the configuration management plan (CMP) and the configuration management function is often given to a separate entity that does not report directly to systems engineering. Historically, such separate entities base their policies on the staircase with feedback paradigm. This is acceptable as long as the CMP is an aid and not a hinderance to the system engineer as well as all personnel involved with design and implementation. Since the CMP becomes of increasing importance to the development process as time proceeds, it is vital for the systems engineer to articulate the particular configuration management needs that exist as a function of the systems engineering paradigm in force.

***Interface Documentation Form and Format***

System and subsystem interfaces represent the major boundaries upon which the systems engineer exercises control. The form and format of interface documentation is developed by the systems engineer. Chapter 9 provides methodologies for this development.

***Test Requirements and Evaluation***

System test requirements are based on and are traceable to system functional requirements. On large projects, actual testing may be carried out by testing organizations separate from implementing organizations. In these cases, generation of test requirements may not be a direct responsibility of systems engineering, but systems engineering will still be called upon to support the execution of tests. This support generally consists of clarifying questions that arise during testing and acceptance of action items in preparation for further test iterations. If there is no external testing organization, then this item should be included in the systems engineering WBS. It is included in the generic WBS at this point for completeness. Chapter 14 covers issues related to testing in more detail.

***Specialty Engineering/Concurrent Engineering***
This is the area of the systems engineering WBS where planning takes place for those special disciplines that may be needed on a part-time basis to support the system design team. Examples of these disciplines include engineering for reliability/availability/maintainability, production engineering, human factors, environment, safety, quality assurance, marketing, customer support, testing, and any other disciplines that may be identified as part of a concurrent engineering approach. If the nature of the mission product and its logistic support require full-time efforts in any of these disciplines, then include such representation on the SDT. If part-time support to the system design team is required, then include these items under the specialty engineering line item.

## EVERYTHING HAS A PLACE

The purpose of the generic WBS is to assist in consideration of everything that has to be done to accomplish system definition, design, implementation, and testing and to provide a guide as to where each activity should be assigned. The structure presented is a sound one that has evolved from and survived many applications. There is, of course, a certain flexibility in how it is used; however, it should not be necessary to deviate very far, if at all.

In deciding where a particular line item belongs in the WBS, there are five simple level 2 questions to ask:

1. Is the item in question a piece of the prime mission product?
2. Is the item in question to be used for purposes of validating, verifying, or testing any part of the mission product?
3. Is the item in question to be used to support development of the mission product or to support operation of the delivered mission product?
4. Is the item in question related to the definition of project management in your environment?
5. Is the item in question related to the definition of systems engineering in your environment?

In this scheme of WBS structuring, everything has a place. If there is any confusion as to where something belongs, identify it with one of these five categories first. Consider the following actual examples.

### Example 1: Radio Towers

Radio towers are to be used to support communications repeater elements. Building towers and building radios are two quite different disciplines. While radio and repeater engineers will certainly supply the radio sets themselves, they typically don't build towers, nor are they even concerned with power or transmission cables that need to go up or down towers. Are the required towers part of the operational support system which comes under ILS or an element of auxiliary equipment under the mission product, or are they a part of the level 3 communications mission product?

Since the radio system link design calls for radios and repeaters to operate at specific heights, the associated towers are an integral part of the level 2 communications mission product. The towers are not a part of the assembly equipment line item because the towers are not to be used by more than one subsystem. The towers are not a part of ILS because they are an integral part of the end item and are directly required for the mission product to meet its functional and design requirements. Thus, at level 4 under communications, would appear the further breakdown of communications elements to include radios, repeaters, cables, tower mounts, and towers as a minimum. The appropriate disciplines to meet these level 4 needs would then be defined and sought as parts of the communications subsystem team.

Alternatively, if the towers were to support not only communications elements, but say, for example, navigation elements as well, then it would be appropriate to assign towers to the assembly equipment line item at level 3, since they would support more than one subsystem. Because we only want one team designing towers, that team would be scheduled in one place under assembly equipment and would be assigned the task of meeting requirements of both subsystems.

### Example 2: Cables

Cables can seem to get complicated very quickly. Here are some guidelines.

Cables between computers and peripherals, for example, clearly are a self-contained part of the computer (or ADP) level 3 mission product item and are a responsibility of that subsystem. Cables servicing or interconnecting different level 3 items should be a part of assembly equipment. Such cables include power distribution from mobile power units in fixed/mobile systems and cables devoted to information flow between subsystems. In complex systems, where there are many cables, it is expedient to devote considerable effort to cable design and layout philosophy. For example, in systems calling for a confusing myriad of cabling, a rational design approach is to standardize pin connections and cable lengths into, say, short-, medium-, and long-length cables. This can be done for both power and data cables if required. In this scheme, there are only three (or six) kinds of cables in inventory that can greatly relieve the replacement problem encountered when a large number of different special cables would otherwise be required. Each cable contains an identical superset of wires required by all subsystems. The female connectors at each subsystem chassis employ only that subset required by that subsystem. These design issues would properly be assigned as a level 4 item under the level 3 assembly equipment line item for the mission product. In complex systems, such as tanks for instance, cabling may warrant elevation to a subsystem in itself. In any case, interfaces should become very clear. The cabling subsystem interface terminates at the male ends of each cable and the communications, navigation, and other interfaces lie at the female receptacles on the unit housings.

**Example 3: Air Conditioning**

Air conditioning is an integral part of environmental control. Its position in your WBS, again, depends on thoughtful review of its use. Air conditioning for an operational command and control center could be assigned as part of *auxiliary equipment* supporting more than one subsystem of the mission product. Alternatively, air conditioning for a development laboratory supporting design could be considered as a part of the facilities line item under development ILS.

While it is evident from these examples that some flexibility in the categorization of specific line items exists, it is important to bear in mind that your hand is not completely free. Your choices are limited by the generic structure itself. Should any particular item not fall neatly into place, then it is a very positive signal that extra thoughtful consideration is required. Slow down and think your way through. Walk away from it and come back as necessary. If you are not sure exactly where each item belongs and the rationale for its placement, then the people who will actually be doing the work can never know.

Before going to a specific example, it is worth taking a moment to emphasize the knowledge gained through the simple diligence of building a sound WBS. Having devoted sufficient time to its thoughtful and thorough construction, considerable insight has been gained for further structuring in eight different specific areas:

1. You can now sketch out top-down system block diagrams. Top-down means you begin by drawing a single block representing the entire mission product, with top-level annotation of inputs and outputs. The next level

of block diagraming exhibits a single block for each mission product line item at level 3. The example given in the next section will follow this process rigorously.
2. As you develop the system block diagram, the subsystems should become evident.
3. As subsystems begin to crystalize, so do the required interfaces between them. If it is obvious at this point that there are overriding considerations for configuring subsystems and top-level interfaces differently, then you may do so (see system partitioning concepts in Chapter 10). Recognize, however, that any changes made at this point will invariably change your WBS. Iteration is, of course, quite acceptable and even expected, but consistency must be maintained.
4. The WBS, by its nature, suggests how you must organize to accomplish your technical goals.
5. The WBS tells you directly what you need to schedule at all levels.
6. The WBS tells you what elements for which you choose to do cost accounting.
7. The WBS helps identify high-risk areas for which effective performance measurement strategies need to be developed.
8. The WBS identifies required disciplines to accomplish your goals—the same disciplines required of the system design team.

It is invariably easier to gain insight into any of the above eight items by first structuring the WBS. The reason is simple. Our primary interests in the system definition process are completeness and consistency.

The generic WBS is the recommended tool for the initial complete definition of what must be done. It is considerably more difficult to initiate such a definition by, say, building a block diagram of something without a structured concept of the complete task at hand. Three-quarters of what must be done (ILS, testing, and systems engineering) are not covered adequately by mission product–oriented block diagrams.

Similarly, it is difficult to schedule something until one knows what to schedule. If the first step in organization is to build schedules, then clearly two things are being attempted at once—that is, the identification of items to be scheduled and scheduling at the same time. It is a lot easier and a lot less dangerous to do one thing at a time.

It is not uncommon to observe practicing systems engineers initiating the process of systems definition somewhere in the middle and then expanding in both ways. A prevalent attitude in the absence of any widely accepted standard is that "everybody is a little bit different and that's the way I do business." The purpose of the generic WBS is to suggest a standard by which a logical structured approach can be executed.

By way of example, consider a system whose function is to automatically gather specific weather data at remote marine sites and periodically transfer current data readings to a central site for further distribution.

The following data is to be gathered: wave height minimum and maximum values in feet, water temperature and surface air temperature in degrees Fahrenheit, surface wind speed in knots, surface wind direction, barometric pressure in inches of mercury, and time of measurements with 1-minute resolution. Time tagged measurements are to be taken every 15 minutes, and capability for storage of up to 16 such measurements shall be provided. The remote units are to be mounted on existing resources, such as buoys, piers, and towers. The system shall be referred to as the Automated Meteorological Data Acquisition and Reporting System (AMDARS).

Level 1 of the system is the AMDARS system. The WBS at level 2 for this system shall generically consist of the mission product, integrated logistics support, testing, project management, and systems engineering items. The following paragraphs discuss the construction of each of these items.

The WBS for the mission product is shown in Table 5-7. Unlike a locomotive, an airplane, or a stand-alone data processing system, the AMDARS consists of a number of identical remote units and a single central unit. The mission products for the remote and central units are different.

In cases like this, there are two basic level 2 segments at the mission product level, an AMDARS remote unit segment, and an AMDARS central segment. The following paragraphs discuss selection of level 2 items, from the generic guidelines and the breaking down of each of the level 2 items to level 3 and level 4 items where appropriate at this stage of activity.

The first step is to review generic level 3 items for applicability to the AMDARS remote unit. Clearly, a suite of sensors is required. Since the units are remote, a communications function is also required. During this first cut, we also determine that a small computer and software will be required to command the sensor suite and manage data communications. We also automatically include auxiliary equipment and assembly equipment. The latter two items are *always* included in the first cut at level 3 of the mission product WBS.

The steps just taken are deceptively simple. For example, if you were constructing a WBS for an earth-orbiting infrared sensing instrument, the sensor itself may not appear in the structure until as far down as level 5, even though sensing is the central mission of the entire system. Table 5-8 shows a partial WBS for this system. This system is more complex than AMDARS. The initial breakdown at level 2 is between the flight and ground segments.

**TABLE 5-7 AMDARS Mission Product WBS**

| Level 1 | Level 2 | Level 3 | Level 4 |
|---|---|---|---|
| AMDARS | | | |
| | Remote unit | | |
| | | Sensors | |
| | | | Wave height |
| | | | Water temperature |
| | | | Air temperature |
| | | | Wind speed |
| | | | Wind direction |
| | | | Barometric pressure |
| | | Communications | |
| | | | Receiver |
| | | | Transmitter |
| | | | Antenna |
| | | Hardware | |
| | | | Computer |
| | | | Sensor interfaces |
| | | | Communications interface |
| | | | Built-in test equipment |
| | | Software | |
| | | | Operating system |
| | | | Executive |
| | | | Communications module |
| | | | Command module |
| | | | Status module |
| | | | Data handling module |
| | | | Built-in test |
| | | Auxiliary equipment | |
| | | | Power |
| | | | Environmental control |
| | | Assembly equipment | |
| | | | Housing |
| | | | Cables |
| | | | Mount(s) |
| AMDARS | | | |
| | Central | | |
| | | Communications | |
| | | | Receiver |
| | | | Transmitter |
| | | | Antenna(s) |
| | | | Data distribution |
| | | | Voice radio |
| | | | Telephone(s) |
| | | Hardware | |
| | | | Computer |
| | | | Communications interfaces |

**TABLE 5-7 AMDARS Mission Product WBS *(continued)***

| Level 1 | Level 2 | Level 3 | Level 4 |
|---|---|---|---|
| | | | Mass storage |
| | | | Printer(s)/plotter(s) |
| | | | Operator's console |
| | | Software | |
| | | | Operating system |
| | | | Programming language |
| | | | Utility packages |
| | | | Communications module |
| | | | Command module |
| | | | Status module |
| | | | Data handling module |
| | | Service system | |
| | | | Vessel |
| | | | Voice radio |
| | | | Instrumentation |
| | | | Support equipment |
| | | Auxiliary equipment | |
| | | | Power |
| | | | Environmental control |
| | | | Work space |
| | | | Storage space |
| | | | Lighting |
| | | Assembly equipment | |
| | | | Cabling |
| | | | Conduits |
| | | | Flooring |
| | | | Connectors |
| | | | Racks, mounts |

The flight segment includes the spacecraft itself, a booster, and so on. The spacecraft is further broken down into a payload, communications, a structure, attitude control, and so on. Finally, under the payload appears the sensor itself.

Thus, the determination of what is placed at level 2 is unquestionably the most important step that one is confronted with in WBS construction. It quite simply determines everything that follows.

We return to our first cut at level 4 for the AMDARS remote unit. Under "Sensors" is listed each sensor type required. Under "Communications" is listed a transmitter, a receiver, and an antenna(s).

It is not clear at this point that a single transceiver and a single antenna

**TABLE 5-8 Infrared Satellite System**

| Level 1 | Level 2 | Level 3 | Level 4 | Level 5 |
|---|---|---|---|---|
| IR System | | | | |
| | Flight segment | | | |
| | | Spacecraft | | |
| | | | Payload | |
| | | | | Sensor |
| | | | | ___ |
| | | | | ___ |
| | | | Communications | |
| | | | Structure | |
| | | | Attitude control | |
| | | | ___ | |
| | | | ___ | |
| | | Booster | | |
| | | Support | | |
| | | ___ | | |
| | | ___ | | |
| | Ground segment | | | |
| | | ___ | | |
| | | | ___ | |
| | | | | ___ |
| | | (Etc.) | | |

might not be used. We are not in a detailed design process at this level of thought. If the mind leaps forward to lower levels of detail at this juncture, it must be stopped and returned to higher-level considerations of functional completeness and consistency at each level of consideration. Depending on one's background disciplines and specific areas of experience, it may be possible to instantly identify significant levels in further detail while working at levels 3 and 4. The analyst will be tugged by a strong urge to show that detail at these levels. Don't yield, but maintain consistent descriptions with regard to detail at each level as you proceed.

Under computer hardware at our first cut, we perceive that a computer will be required at the remote unit. Hardware interfaces between the computer and the sensors and between the computer and the communications element shall also be needed. *Built-in test equipment* (BITE) is also included for completeness. The need for BITE will be determined by system functional requirements.

Software at level 4 includes any and all items we believe may be required at this point in the early thinking. These include an operating system, an executive (or control) module, and application modules for handling communications, commands, unit status, and data. Note that some of these modules may change or be modified as to their functions as a result of later review by the system design team. For example, the functions now associated with a data handling module may later be absorbed by other modules—that is, an error detection function may easily be included in the communications module, or message decoding may be handled by the executive module. *Built-in test (BIT) software* is also included. The important issue is that the SDT is presented with as functionally complete and consistent picture as possible. The integrity of the structure is an important part of the systems engineer's leadership. The systems engineer will not do the detailed design—he or she will lead it.

Auxiliary equipment for the remote unit will certainly include power. Environmental control for one or more of the sensors may also be required and is included at this time for the sake of completeness. Issues of this type will be driven by functional requirements and the design activity that responds to those requirements. It is far easier to simply remove a line item than to find it necessary to add one later when design concepts and interfaces are being established. Errors of omission are far more serious than the inclusion of items that prove to be unneeded. Removing items from an established structure is inherently easier than finding that a needed addition requires alteration of the structure.

Finally, we include under assembly equipment the housing for the remote unit and necessary cables and mounts.

The AMDARS central unit is the same as the remote unit at level 3, with the exception of the need for a sensor suite and a service system. There are major differences at level 4.

In addition to a receiver, a transmitter, and an antenna(s), the communications element for the central unit includes facilities for data distribution to system users and voice communications consisting of telephone and radio communications. Again, restrict your enthusiasm at level 3 and maintain consistency of description throughout.

Hardware includes a computer, communications interfaces, a data storage medium, a printer(s) and a plotter(s), and an operator's console.

Software packages consist of an operating system, a language for application software, utility packages, and modules for communications, commanding, status handling (running BIT), and data handling. Any of these categories may change as a result of eventual response to functional requirements or discussion with your design team. Building a WBS at this point in the systems engineering process is not a design function. The primary purpose at this stage is to provide a mechanism to methodically organize.

It is also perceived at this level that a subsystem will be required to install and service remote units. This will consist of a vessel, voice radio communications with central, marine instrumentation, and support equipment for safety and for installation and removal of remote units.

Auxiliary equipment at central includes power and environmental control units. Also included are work and storage spaces (desks, chairs, cabinets, etc.) that are directly related to the functions to be carried out at central. Items such as washrooms, drinking fountains, showers, safety gear, and so on are to be included in the facilities item under ILS.

Finally, assembly equipment at central includes all necessary items to install and connect other level 2 items, such as cables, conduits, special flooring not included under facilities, connectors, racks, mounts, and so forth.

At this level of preliminary organization, the WBS is completed by appending level 2 and level 3 items for integrated logistics support, system testing, project management, and system engineering exactly as they are listed in Tables 5-2 through 5-6.

At your discretion, you may wish to enter entries at level 4 under ILS to complete your preliminary thinking on where specific items may fit into the total picture. For example, the central segment clearly requires a structure at level 4 under the level 3 "Facilities" entry. It is not important at this juncture whether the structure will be a stand-alone building or part of an existing or planned building. The thinking is not limited to those items that need to be designed only, but also includes items to be procured, leased, inherited, and so on.

With a preliminary WBS completed, we have made constructive strides, based on a knowledge of the user needs statement, toward understanding what the complete system consists of, what its boundaries are, and how its elements will basically interact. The preliminary work is ready for assessment by the SDT.

The analyst is also prepared at this time to sketch initial system block diagrams based on the WBS. Figure 5-1 presents the system at level 1. System inputs and outputs are annotated at the highest level.

Figure 5-2 presents the system at level 2 with appropriate functional annotation at a next level of detail.

Figure 5-3 shows a block diagram at level 3 for the remote unit segment. It is immediately evident that the sensor suite has interfaces with the data

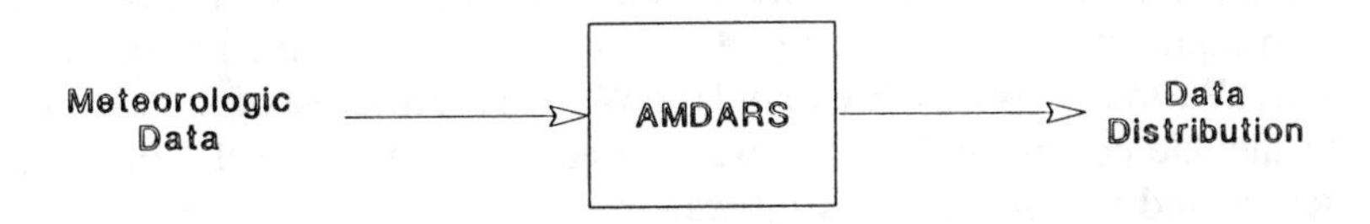


**FIGURE 5-1.** AMDARS block diagram—level 1.

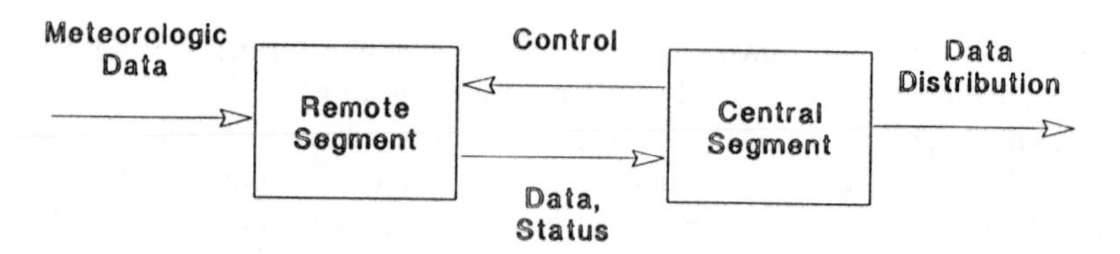


**FIGURE 5-2.** AMDARS block diagram—level 2.

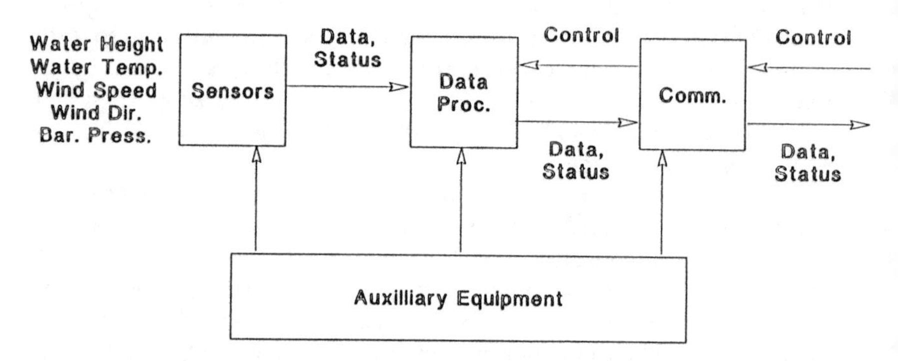


**FIGURE 5-3.** AMDARS remote segment block diagram—level 3.

processing hardware and that the hardware has interfaces with the communications subsystem. A similar diagram is easily constructed for the central segment.

The system block diagram example emphasizes four important points:

1. The functional block diagrams faithfully follow the WBS at each level.
2. Functional interfaces between subsystems are beginning to clearly emerge.
3. Iterations to the WBS may begin to take place. Note, for example, the new choice of words "data processing hardware" in Figure 5-3. This represents a refinement of the original term "hardware" in the remote unit segment WBS. Note also that, during the original thinking that went into the WBS, communications and sensor interface items were included under hardware. We may now wish to expand the WBS in the interest of completeness to include corresponding hardware interface items at level 4 under the sensor and communications subsystems.
4. Note that not all of the level 4 remote unit segment items are included in the functional block diagram of Figure 5-3. Not only are the level 4 mission

product items of software and assembly equipment not explicitly shown, but the level 2 items of ILS and systems engineering that also provide system products are also not included.

This exercise should clarify the importance of initiating system definition with construction of the WBS. Functional block diagrams do not provide a complete view of the entire system. Similarly, a complete understanding of what needs to be scheduled can only be gained by viewing the total system as represented by the WBS. Finally, the WBS must be a living document that is constantly updated and improved as the design and implementation process matures.

# 6

# The Systems Engineering Management Plan

The systems engineering management plan (SEMP) is the top-level technical management plan that states how the goals of systems engineering are to be met for a specific project.

Construction of a comprehensive SEMP represents a valuable opportunity for the systems engineer to define the technical structures and processes to be employed in the complete execution of systems engineering responsibilities. The plan must be consistent with and supportive of higher-level management plans.

The document should not be overbearing or lengthy, but should consist of a series of succinct paragraphs that state the strategies to be employed in key areas of technical management tailored to the project at hand.

Suggested topics to be included in the SEMP are listed in Table 6-1. Detailed discussion of each of the topics is provided in the referenced chapters in the table. The following paragraphs briefly describe the scope and content for the treatment of each topic in the SEMP.

## ORGANIZATION

This paragraph includes brief text supporting an organization chart showing the reporting relationships between the project manager, project staff, systems engineer, cognizant engineers, concurrent engineering teams, and system user organization. If appropriate to the project, the positions of project engineer and software systems engineer are also included. The project organization, discussed in Chapter 15, is of critical importance in that many system failures can be directly attributed to structures that are incapable of properly executing systems engineering functions.

**TABLE 6-1 Suggested Topics for the Systems Engineering Management Plan**

| Topic | Chapter Reference |
|---|---|
| 1. Organization | 15 |
| 2. Systems engineering paradigm selection | 2 |
| 3. The system design team | 7 |
| 4. Requirements definition | 4, 10 |
| 5. Prioritized competing design characteristics | 12 |
| 6. Margin management | 15 |
| 7. Options analysis | 12 |
| 8. Logistics support | 11 |
| 9. Configuration management | 8 |
| 10. Performance measurement | 15 |
| 11. Risk management | 12 |
| 12. Test and evaluation | 14 |
| 13. Schedules | 15 |

## SYSTEMS ENGINEERING PARADIGM SELECTION

This paragraph contains a brief statement of the rationale for selecting either the staircase with feedback, early prototype, spiral, or rapid development models, or a combination thereof.

## THE SYSTEM DESIGN TEAM

This specifies the permanent membership of the SDT by title and specifies temporary membership by discipline that will be called upon to support the SDT, including required disciplines in specialty and concurrent engineering. The paragraph also states how often the SDT will meet, gives meeting location rotations (implementor, user, contractors), if appropriate, and establishes the roles and responsibilities of the SDT.

## REQUIREMENTS DEFINITION

This states the strategy planned for formulating user needs (if required), system functional requirements, and system allocation of requirements (partitioning). The strategy includes who will take part in each function, responsibilities for each function, and the rationale for system partitioning. The strategy is intimately related to the choice of the systems engineering paradigm for the project.

## PRIORITIZED COMPETING DESIGN CHARACTERISTICS

This lists the project prioritized competing design characteristics (PCDCs), defines each characteristic, and describes how they will be used as top-level criteria for system design trade-offs and related technical decisions.

## MARGIN MANAGEMENT

This identifies candidate scarce commodities for which margins will be established and maintained. It presents margin management philosophies and methodologies to be established at SRR, PDR, CDR, and other points in the development cycle that are deemed appropriate. The philosophy for margin management is closely structured by the selected systems engineering paradigm.

## OPTIONS ANALYSIS

This identifies points in the development cycle where a need for resolution of requirements and design issues is anticipated, if required. It describes appropriate methodologies, such as early prototyping, use of static or dynamic models, breadboarding, use of a series of operational deliveries, and so forth, that will be used to gain confidence in proposed options. The structure of the options analysis strategy is influenced by the systems engineering paradigm in use.

## LOGISTICS SUPPORT

This describes the top-level strategy for operational logistics support to include discussion of each major logistics support item that needs to be addressed in the design of the system for the project at hand. Depending on the project, major items may or may not include concepts for supply support, test equipment, technical data packages, transportation and handling, facilities, and personnel and training. A succinct description of the operational maintenance philosophy is included, describing how the system will be maintained and by whom. A similar description of major logistics considerations is included to support the development of the system, as required.

## CONFIGURATION MANAGEMENT

This states the basic approach to configuration management as it will be applied to the project. It clarifies the extent and timing of control to be executed as a function of the systems engineering paradigm to be used. It defines baseline freeze points for requirements, system specifications, subsystems designs, and other baselines, as required. It also defines how a single set or multiple set of repeated baselines will be established in accordance with the chosen paradigm.

## PERFORMANCE MEASUREMENT

This defines the tools and processes by which the systems engineer plans to maintain visibility over the development process. The discussion specifies which tools, such as Gantt charts, PERT charts, low-level schedule monitoring, reviews, walkthroughs, early testing, earned value determination, and so on, will be employed for the technical monitoring of system development.

## RISK MANAGEMENT

This identifies any anticipated potential high-risk issues likely to be encountered throughout the development cycle. This discusses the risk management approach and timing to be employed, which is basically driven by the systems engineering paradigm in use. This defines the top-level technical backup design and implementation strategies to be used in the event of failure of initial development efforts.

## TEST AND EVALUATION

This discusses the top-level sequence strategy for system integration testing, system level, and acceptance testing as a minimum. Other testing strategies may involve the need for pre-ship and post-ship testing. This paragraph also states when and by whom the test plans and procedures and detailed test plans will be developed as a function of the systems engineering paradigm employed.

## SCHEDULES

This presents a top-level schedule on a single page for the major systems engineering deliverables in a manner consistent with the adapted systems engineering paradigm for the project.

## DEPARTMENT OF DEFENSE SEMP FORMAT

Defense systems routinely involve heavy participation of contractors. The Defense Systems Management College (DSMC), along with the United

States Air Force MIL-STD-499A on Engineering Management, calls for a specific format for the SEMP. The document is organized into three parts. The following descriptions of each part are taken directly from the DSMC's *System Engineering Management Guide*:

Part I: Technical Program Planning and Control—Identifies organizational responsibilities and authority for the contractor's system engineering management, including control of subcontracted engineering, verification, configuration management, and document management, as well as plans and schedules for design and technical program reviews.

Part II: Systems Engineering Process—Describes the process used in allocating and defining requirements and their documentation for the program. This part also explains the contractor's intended strategy for generating multiple alternative designs at each development level, and the trade-off results which trigger iteration of the design process.

Part III: Engineering Specialty Integration—Defines how the engineering specialties of reliability, maintainability, human engineering, safety, logistics, producibility, and other areas are integrated into the mainstream design effort.

The DOD version provides a format for the contractor's response to the requirement to generate a SEMP. The basic role of the document, however, is still to identify and control the overall systems engineering process. The items listed in Table 6-1 provide an extended guide to the content of the SEMP.

In the generic systems engineering process, the SEMP is generated by the systems engineer immediately after the SDT is formed— that is, early in the overall process. The SEMP is then reviewed by the SDT, iterated as required, and submitted to project management for approval.

# 7

# The System Design Team

The SDT is the platform from which systems engineers organize and lead the majority of their work. The SDT is the group through which the systems engineer accomplishes the production of requirements at all levels, the system design, fabrication, test, installation, and final approval.

It is highly desirable that the systems engineer have a major role in determining both the function and the specific makeup of the SDT. Unfortunately, this is not always possible as management often has predetermined notions regarding the boundaries of the systems engineering role and often selects key players prior to selecting the systems engineer. In these cases, careful review of the team's function and makeup is called for and appropriate recommendations for modifications are in order. This is particularly true when users are inadequately represented on the design team.

## DESIGN TEAM FUNCTIONS

Table 7-1 lists the major functions of the SDT. Most items are self-explanatory and are covered in appropriate sections of this book.

The systems engineering management plan includes a definition of system PCDCs commodities for margin management, and other considerations under the direct purview of the SDT.

The system-level portions of the functional requirements and system specifications are generated by systems engineering, and the functional area, or subsystem, sections are generated by the appropriate subsystem cognizant engineer team members.

Selecting design tools and aids involves the choice of such items as an appropriate *program design language,* automated design aids and strategies for using structured analysis, structured design, unit folders, and other concepts useful in structuring and controlling design activities.

**TABLE 7-1 Principal Functions of the System Design Team**

1. System engineering management plan iteration and concurrence.
2. Work breakdown structure iteration and concurrence.
3. Development of system functional requirements, system specifications, system test requirements, and any required subsystem documentation.
4. The making and monitoring of detailed schedules.
5. Identification and resolution of system-level issues.
6. Options analysis.
7. Margin management.
8. Interface definition.
9. Configuration management support.
10. Selection of design tools, aids, and strategies.
11. Project management support.

Project management support consists of periodic support in scheduling and costing, assistance in structuring formal reviews and their presentation, and attendance at informal meetings.

## DESIGN TEAM MEMBERSHIP

The systems engineer is the chairperson of the SDT. The following functions, as a minimum, should be represented on any design team on a permanent basis:

1. Systems engineering;
2. Cognizance for each level 3 WBS mission product item;
3. Software systems engineering;
4. ILS engineering;
5. Test engineering;
6. User representation; and
7. Other specialty and concurrent engineering team members, as required.

Item 1 is represented by the systems engineer, and, on larger projects, typically a lead member of the systems engineering staff is also included.

On larger projects, a single COGE may be warranted for each of the level 3 mission product items. On software-intensive projects, subsystem software engineers may be included. Alternatively, the *system software engineer* may run his or her own *software system design team* (SSDT) and report on software activities to the SDT.

On smaller projects, a single engineer may have cognizance over more than one mission product item, or, for example, ILS may be absorbed into the systems engineering role. All items, however, should be represented.

ILS and test engineering representation ensures that important design impacts driven by these functions are not overlooked.

Item 6 above cannot be overstressed. *A major cause of system failures lies in constructing systems that do not do what the customer wanted. You cannot run a successful design team without constant customer representation.*

An isolated implementation team that is designing and fabricating a system of any reasonable complexity based solely on the existence of top-level functional requirements, without the presence of the customer, is extremely dangerous. Holding baseline reviews for the customer, while necessary, is typically not enough. Constant representation of the customer's view is crucial, particularly when developing systems that are meant to work problems that have not been specifically encountered before, as well as systems simply designed to provide major improvements over existing systems. In these settings, requirements can change as insights are gained through the development process. Engineers are notorious for solving problems with the most expedient "engineering" solutions, which may or may not be customer oriented. When left alone to cope with the myriad of issues that arise throughout the development cycle, the systems engineer can absolutely count on going astray of user needs in the absence of adequate user feedback.

This is not to suggest that the customers can constantly change their minds with regard to fundamental direction. Baseline commitments should be seriously maintained. Rather, the customer is there to maintain the development team's touch with reality. The customer is there to guard against the building of a system that cannot be used or that is not wanted. The fact that exactly this happens in an alarming number of instances is warning in itself.

Most vendors, unfortunately, don't like continual customer participation in their design activities and decisions. This mind set, often driven by proprietary reasoning or a simple "dirty linen" defense, is perfectly understandable. The history of systems engineering clearly indicates the inherent dangers of development in isolation. When change is truly required, it must be identified and accommodated in a timely manner. The probability of success is much greater when the customer is represented on the SDT.

Additional membership oriented toward speciality and concurrent engineering on a full- or part-time basis may consist of the following functions:

7a. Production and value engineering;
  b. Reliability/availability/maintainability (RAM);
  c. Human factors engineering;
  d. Safety engineering;
  e. Quality assurance;

f. Environmental engineering;
g. Finance;
h. Marketing;
i. Customer service;
j. Suppliers; and
k. Other specialties, as required.

The representation of a diversity of interests is clearly evident. The representations shown here are suggested ones. The point is that the SDT must include any and all interests that may effect product development and use. It is of little value, however, to establish such representation if the SDT does not have sufficient authority to carry out its responsibilities. It is a matter of fact that successful execution of the concepts of concurrent engineering often requires realignment of authority—that is, reorganization. Conventional nonconcurrent engineering structures can seldom accommodate the new authority required by a concurrent team (see Chapter 15).

Generally, the size of the project, the complexity of design issues, and the degree of concurrent engineering employed dictates the number of actual people required to fulfill these roles.

As a rule, a permanent SDT membership of more than ten people should be avoided. Meetings with larger groups tend to impede progress. This means that on big projects the permanent membership must be made up of people who are not only technically competent, but are also capable of delegating work and providing leadership to those reporting to them. On the other hand, if there are only 13 people on the project, there is good reason to include them all.

Specialty engineering includes such items as RAM, value engineering, human factors, environment, safety, Q.A., and so on. These disciplines may not always be required on a full-time basis. Part-time members are perfectly acceptable as long as their need at crucial points of requirements setting, design, and test is understood and not neglected. This sensitivity requires a constant vigil.

In this light, systems engineers must be knowledgeable enough to know what they don't know. It is interesting that, if you ask engineers to perform brain surgery, they will generally admit that they don't know the first thing about it. But if your subject is within the bailiwick of the wide diversity of the many engineering disciplines, there is a very definite tendency for engineers to "know it all." I have seen extremely competent managers in every other respect completely underestimate the importance and subtleties of ILS, for example, or not know the difference between reliability and availability. And, of course, everyone knows everything there is to know about systems engineering.

The difference is that engineers know they can't perform brain surgery, but many of them simply don't know what they don't know about the many facets of "engineering."

The familiar term "interdisciplinary" is used to refer to the mixing of totally different fields of endeavor. Today the single field of engineering is so diverse that it has even become interdisciplinary within subfields. Consider the vast differences between computer science professionals specializing in software development, VLSI fabrication, bus design, quality control, training, and so forth. It is not reasonable to expect a single person to know "everything." It is, in fact, impossible for a single person to know everything. It is nonetheless true that the systems engineers must *know what they don't know.*

The lesson is that we must try very hard to understand the boundaries of our knowledge. Systems engineers, in particular, must take the time to learn enough about every technical discipline that affects them in order to propitiously recognize when to get help. There is nothing wrong in this. The systems engineer is not expected to know everything, but he or she is most certainly expected to be aware of everything.

## RUNNING THE TEAM

Design team meetings should take place once a week. The selection of the particular day can have advantages.

Fridays have the disadvantage that people tend to lose steam over the weekend. People also generally go through a transitional start-up period on Mondays and gain momentum as the week progresses. If they have received an Action Item assignment during a Friday meeting, they are more likely to lose some of the crisp detail of what must be done between then and Monday.

Providing updates to your directions on Wednesdays enhances the probability that work on actions will begin during the latter part of the week when productivity is still on the rise.

Avoiding Monday also provides the systems engineer with time to prepare a productive setting for the meeting. It is useful to informally move among the team members on Mondays and Tuesdays and make a point of discussing issues that may be controversial or even routine, sometimes one on one with specific individuals and other times with small groups. If you sense that two people have opposing views or approaches, you may use this opportunity to discuss the issue individually with each to understand their viewpoints in an effort to define a middle road that does not violate the goals of systems engineering. You may wish to go back and forth between them several times or to get them together to resolve the issue. A basic goal in this politicking is to prepare for a smooth meeting.

This is not to suggest that issues should not freely be raised and discussed during design team meetings. The goal, rather, is to minimize outright conflict and create a general meeting atmosphere of civility and cooperation. This is done by identifying and resolving serious conflicts outside of the meeting. Design issues can be the source of heated discussions, some technically oriented and many simply personality oriented. After some years of conducting design team meetings that from time to time turned into roaring debates, I successfully adapted the approach of trying to anticipate delicate design and personality issues in advance and tackling those issues outside of the meeting forum. There is an important psychology to this. Civility breeds civility. It sets the tone of your command. Open refusal to adapt your proposed course of action can rapidly diminish your credibility as a leader. It is quite feasible to reduce the probability of spontaneous rejection at the design team meetings by minimizing surprise. The Monday and Tuesday preparatory mini-meetings provide a significant opportunity for the systems engineers to obtain support for the direction and assignments they intend to put forth in the actual design team meeting. By anticipating and resolving conflict, surprise and dissension can be significantly reduced.

In these pre-meeting discussions, it is not wise initially to dictate. A more productive approach is to make the immediate goals clear and then ask, "How do you think we should go about this?"

It is also useful to discuss the progress of assigned action items. This affords the opportunity to discuss in advance any implications for other team members that may be affected, as well as any of their line supervisors, as appropriate, that like to "keep up." It is expedient to make such "rounds" on Mondays and Tuesdays, as needed, which allows for the preparation of the terrain for a well-organized meeting in which direction occurs smoothly because the principal parties are informed.

People respond very well to this treatment, as it is always a compliment to be consulted in private or in small groups in advance. This kind of preparation typically results in nods of agreement as opposed to surprise or shock. Because most people are, at best, tentative to embrace any new idea on the spur of the moment, surprise and the often lengthy discussion that follows is generally to be avoided. An atmosphere of smooth cooperation is very important to attain early in the game, as this tone rapidly becomes a permanent characteristic of the group's behavior. Normally antagonistic people, real shooters from the hip, become highly civilized in such cultivated settings. Major "public" disruptions in a design team meeting can drastically erode your credibility, both technically and as a leader before the group. If they continue over any reasonable period of time, your effectiveness will rapidly degrade. The intent here is not to mesmerize people. On the contrary, the aim is to provide time for thoughtful consideration of identified issues.

In this scheme, much of the effort on the part of the systems engineer to direct the overall process is expended outside the SDT meeting.

Each meeting should begin with a review of level 4 and/or level 5 schedules. Expect revisions to be called for. If weeks go by and everything remains "on schedule," it probably means that the schedules are not detailed enough and there is a risk of loss of visibility. It is also possible that you are not being told everything that is happening.

Alternatively, if all line items are constantly changing, it is likely that the schedules are too detailed. In discussing schedules, address the time frame of 1 to 4 weeks around the present date in sufficient detail. If a team member asks for another week or two to complete a specific line item, be ready to give it, but discuss the impacts on other line items and formulate approaches to maintaining the schedule for the item end date if needed. Changes in line item end dates on a given schedule may affect that item on the next level up. The working schedules under discussion at the design team level should be below the level of schedules you routinely present to your own management. Thus, you have your own built-in pad and flexibility on week-by-week schedules, but you must stress the importance of meeting end dates that are reflected in the higher-level schedules that management is holding you responsible for.

In addressing the work to be done in the following week and month, be sure this is the work that is actually taking place. If it is not, add line items as required and adjust other line items if needed. Maintain visibility. Strive to understand the reality of what is happening without issuing reprimands. Ask if the level of confidence in the existing schedule with respect to the items and their timing is still good. Stress from time to time that this schedule represents your commitment as professionals. Be sure the schedule reflects the real work and that everyone is professionally comfortable with it. Spend the time to review every item. Ask your questions and then be quiet.

Some years ago, my boss came to me after sitting in on one of my design team meetings.

"Do you realize," he said to me, "that your meeting lasted one and a half hours and you were talking for over one full hour of that time?"

"Was I?" I said, slightly amazed.

"They all know what you're thinking," he said, "but do you know what's on their minds?"

He helped me realize that you don't learn by talking.

Draw your team members out as much as possible. Listen. If you sense that a serious slippage is possible, beyond the limits of your own pad, ask their advice on how to recover. Make every effort to implement their plan for recovery. Move other line items and make temporary shifts in personnel assignments. Involve the whole team in attacking the problem. Stress that the discovery of a problem is a positive event. A constant goal in scheduling is to discover problems with the schedule as early as possible—weeks, even months in advance. This is, in fact, why schedules are built. Clearly, the earlier a scheduling problem is revealed, the greater the options are for its resolution.

If action items are called for, be sure to pinpoint the exact course of action to be followed. If the need for the action item emerged during the meeting and not as a result of your pre-meeting preparations, do this interactively by proposing approaches or simply raising the question, "What is the best way for us to do this?"

Never complete your discussion of an action item without a clear understanding of what individual is responsible for its closure and a target date for closure. Always identify a single individual as being responsible for action item resolution, even though that individual may lead a subgroup to perform the actual work. Assignment of an action item to an individual need not always mean that that specific person will be doing any or all of the work. He or she may complete all of the work, do part of the work with help from others, or totally delegate the work to be done to others. The important point is that more than one person cannot be responsible for an action item closure. It is too confusing to say, "Charlie and Ed will take care of that one." Either Charlie does or Ed does, and gets help from others as appropriate. Also, take the opportunity, when appropriate, to assign action items to yourself. Be a team player.

Throughout this entire process, constantly look for opportunities to pass out compliments to individuals before the entire team. This need not be overdone, but such opportunities should be taken as they logically present themselves. Over time, try to do this for everyone in a balanced fashion and in an honest and sincere way.

After a thorough discussion of schedules, it is useful to summarize the current state of schedule items. For each issue that has arisen, you must decide whether each will be resolved through the normal course of execution of schedule line items or whether special action outside of the existing schedule is called for. If special action is required, review each action item, including who is responsible for closure, any agreed-upon support, and the due date.

In the final phase of the meeting, go around the table left to right and ask each person if they have anything else to bring up. Simply asking the group at large if "anyone has anything else" is not sufficient to draw out the information you need.

For example, your conversation might go like this:

"John, do you have anything else?" Then be quiet and wait. If John says, "No," then say something related to his efforts, anything.

"You've been holding the schedule pretty well. Do you still feel good about what we've laid out for the next few weeks?"

John says, "Yep."

Keep up the pursuit. "We don't have a lot of flexibility, but we have some. We can make some changes."

Often a respondent will offer an inner thought that would not normally be illicited had a more impersonal or cursory approach been taken. At this point John might say, "There is one thing I might bring up" or "Line item six might be a little tight," and so on. This is a good result. There *is* something on John's mind, and it is very important for you to make this discovery and to get it out. Remember, the name of the game is to maintain visibility, and you must communicate in order to achieve that goal.

You don't have to do this in great depth with every team member at every meeting. Some people need very little prompting to talk incessantly. Others tend to remain quiet over long periods of time. These are the ones you need to encourage to participate. Often they know more about what needs to be done than anyone else, including you. Use your judgement, but do try to probe everyone over a period of time. Much of this has to do with who's involved with what is currently going on and your own confidence in how the work is progressing. But definitely address each person around the table sequentially at every meeting, and pause noticeably to give them time to respond.

Your design team is your lifeblood. To every extent that you lose touch with your design team, to that same extent do you lose touch with reality. Your ability to communicate is imperative!

Just before concluding the meeting, it is useful to review each agreement and action item assigned. Do this in your own words to establish a final understanding of the meeting's outcome and the wording of the results to be used in the meeting minutes.

The major characteristics of the design team minutes are:

- Actions taken;
- Agreements made;
- Resulting action items:
  - *What* is to be done?
  - *When* is it to be done?
  - *What individual* is responsible?

This will give you an audit trail of the evolution of your engineering decisions and design. The minutes reflect that the meeting has concentrated on actions taken, agreements made, and action items identified and assigned.

An example of design team minutes to be issued after every weekly meeting is given in Figure 7-1. You may designate a secretary to generate your minutes or write them yourself. The generation of minutes by the systems engineer has specific advantages for at least three reasons.

| | |
|---|---|
| TO: | Attendees and Distribution |
| FROM: | (Your Name) |
| SUBJECT: | Minutes of Project X Design Team Meeting of 6/12/91 |
| ATTENDEES: | J. Brown, N. Green, J. Grey, M. Maroon, R. Opal, R. Purple, G. Red, (your name) |
| NEXT MEETING: | Wednesday, June 19, 1991, 10:00 AM, Bldg. 204, Room 128 |

Development schedules were reviewed. All items remain on schedule with the following exceptions; 1) Procurement for the X-2000 terminal has been delayed for an estimated two weeks due to vendor backlog. 2) The scheduling of code of software module FOO will be interchanged with software module BOO which is not dependent on the X-2000. 3) A line item for software conversion to accomodate system queueing for the new plotter will be added to the level six schedule. There are no schedule slippages anticipated at level 4 as a result of these changes.

G. Red reported on results of disc access simulation studies. A recommendation to upgrade the operational system disc unit to model 502 was made as a result of these studies. The recommendation was accepted by the team. Action item No. 23 is closed.

It was agreed to extend the current development software maintenance contract for one additional year.

An action item (No. 31) was assigned to N. Green to develop strategies for reduction of protocol overheads for the Boston communications link by 10%. Action item 31 is due on July 10, 1991.

DISTRIBUTION: M. Black<br>L. Blue<br>K. Cordovan<br>S. Hue

**FIGURE 7-1.** Sample system design team minutes.

First, it provides an opportunity to mention peoples' names as often as possible and include everyone's name over a period of time. The team members' supervisors generally read the minutes, and it is a subtle way to give everyone points.

Second, it is practical to control the honesty of what is reported. When good things happen, they should be conveyed to management. When a problem arises, it should also be noted, along with a plan for action. If a problem persists, management then has some visibility. Surprises are almost always undesirable for everybody.

Third, the minutes serve as a valuable audit trail for the development of design and the rationale for that evolution.

The text portion of minutes should rarely run in excess of a single page. Each item covered represents a point worth recording in the context of the overall flow of development. The issues are technical. Administrative issues are seldom included, these being covered in staff and project meetings. Results of action items and special studies are attached as appendices for design team distribution, but need not be widely distributed to management.

The body includes a list of attendees, announcement of the next meeting location, date, and time, a series of paragraphs summarizing salient issues covered, and a distribution list. Attached to the text portion is a copy of the current level 5 schedule with updates.

Also attached is a list of outstanding hardware and software action items assigned at the team meetings. Separate lists of failure reports and change requests that result from formal test exercises or from experience with previous operational deliveries should also be attached as required. These documents include an identification number for each action item, FR or CR, the name of the initiator, the date of initiation, a one-line summary of the issue, the assignee for action, a due date, and a comment field. An example of an FR list is given in Figure 7-2. Closure of items is noted in the comment field, and the item is then removed from subsequent listings. Thus, it is evident from Figure 7-2 that items 1, 3, 5, 6, 8, 9, and 11 were closed in meetings prior to 6/21/91. The log of SDT minutes contains a complete history of the opening and closure of items.

Separate listings for action items and CRs follow the same format as shown in Figure 7-2. These reports must be integrated with the configuration control, auditing, and status accounting mechanisms discussed in Chapter 8, "Configuration Management."

In large organizations, it can often be difficult for an independent configuration management function to provide complete updated information on the exact current status of FRs and CRs to adequately support weekly design team meetings. When delays of this nature occur, systems engineers may wish to produce an informal weekly report on their own PCs for use at the design team meetings, following the format in Figure 7-2. While this informal approach may provide more timely information, it should be remembered that the formal status accounting and reporting process is the one that is visible to management. This is true whether the formal mechanism is under the direct control of the systems engineer or under the control of a separate organization. If such informal weekly tracking is expedient, it should also serve as a weekly update mechanism to any more formal process in use.

Finally, a primary goal of the systems engineer is to gain a consensus among SDT members that is consistent with the current and long-term goals

PROJECT X FAILURE REPORTS AS OF 6/12/91

| FR | INITIATOR/ OPEN DATE | PROBLEM | ASSIGNED TO/ DUE DATE | COMMENTS |
|---|---|---|---|---|
| 2 | J. Oak 3/12/91 | Text Editor | N. Green 6/22/91 | |
| 4 | T. Elm 3/18/91 | CPU-4 RAM | R. Opal 6/5/91 | Vendor shipment expected 7/1/91 |
| 7 | A. Maple 4/16/91 | NAV Module formats | G. Red 6/28/91 | |
| 10 | N. Palm 5/23/91 | FFT module | J. Grey 7/1/91 | Speed enhancement |
| 12 | J. Oak 6/2/91 | User Terminal-6 | R. Opal 6/12/91 | CLOSED-Notify CM |

**FIGURE 7-2.** Sample failure reports listing (same format for action items and CRs).

of systems engineering. Consensus is required with regard to all SDT activities, including WBS construction, schedule making, interface definition, selection of design criteria and priorities, development and documentation of requirements and specifications, identification and execution of special studies and options analysis, action item assignments, choice of design tools and aids, assignment of failure reports, assessment of user feedback and lien corrections, and all other pertinent business of the SDT that the systems engineer deems mandatory.

Consensus is best obtained by clearly asking for it. For example, when discussion reaches a point where a conclusion seems imminent, it is wise to reiterate the conclusion in summary by stating, "We all agree, then, that ..." and to probe those members of the team whose disciplines are most affected by asking, "Does that sound reasonable to you, Ed?"

If discussion is not leading toward a consensus, the systems engineer should try immediately to determine whether consensus can or cannot be reached in the framework of that particular meeting.

If it is felt that consensus is possible in a reasonable amount of time, it can usually be extracted by guiding the discussion through asking questions. The questions should be specific and directed toward the consequences of unacceptable action(s) under discussion. If you can ask the right questions in the right sequence, other team members often offer comments that bring the discussion around to the conclusion that you wanted in the first place. It is always better for someone else to come up with the solution you want. You must, of course, be right to begin with and be able to pose a structured sequence of questions that lead to a logical conclusion.

If it is felt that consensus on an issue cannot be reached in the time frame of the meeting, either because of uncertainty in what should actually be done or because of lengthy discussion, an action item should be assigned to an appropriate individual to study the issue and recommend a course of action at the next meeting. This not only gives the systems engineer an opportunity to think upon the matter further, but it also provides an opportunity to discuss the issue in private with the assignee over the next week(s).

Ideally, SDT meetings should not last much more than 1 hour, although they may on occasion last up to 2 hours. Within this time frame, if consensus on a given issue is not reached within 10 to 15 minutes after sufficient discussion has taken place, it is likely that it will not be reached on that particular date.

The SDT meeting is not a working meeting. Rather, it is geared toward reviewing status, identifying system issues, and maintaining team coordination and communication. Issues that are not resolved in a timely manner are best delegated and addressed outside the design team meeting in routine working meetings that focus on those single issues.

The SDT is the medium through which you set your tone as a manager. There is a difference between being a boss and being a leader. Bosses give orders. Leaders do just that—lead. They lead by identifying and drawing out the best qualities in people and focusing those assets on problem identification and solution. Bosses generate fear. Leaders do not generate fear—they generate cooperation. Leadership entails the ability to listen, to involve people who know more detail than you do, and to allow them to achieve pride through contribution and workmanship. The good leader commands respect through technical thoroughness, gives respect to those more knowledgeable, understands and sticks to the generic systems engineering process, and builds human beings by focusing their best abilities toward a common purpose.

# 8

# Configuration Management

While the evolution of systems engineering discussed in Chapter 2 has given rise to the need for thoughtful modifications to traditional configuration management (CM methods), it remains true that CM has always been (and will continue to be) a vital tool necessary for maintenance of control over system requirements, design, and development. This chapter reviews the fundamentals of classic CM. In the staircase with feedback paradigm, it is applied as presented here. When the spiral model or the rapid development model is used, modifications are required to serve the needs of flexibility that these paradigms call for.

In the early prototyping and spiral models, the modifications consist of an extension of the early, more relaxed phases of classical CM with regard to baseline freezing until requirements and designs are fully developed.

The rapid development model calls for the repeated application of classical CM with each system delivery. Further, a less strict application for the first deliveries and progressively more stringent application with later deliveries is called for.

## CONFIGURATION MANAGEMENT—A DEFINITION

Configuration management is the discipline of identifying selected elements of a system at discrete points in time for the purposes of systematically controlling changes to the configuration and maintaining the integrity and traceability of the configuration throughout the system life cycle.

There is general agreement in management literature that CM consists of four major elements. They are:

1. Identification;
2. Configuration control;
3. Auditing;
4. Status accounting.

Tables 8-1 through 8-4 summarize the definitions and basic questions at hand in dealing with these four components. The following paragraphs discuss these in order.

## IDENTIFICATION

Configuration identification is the careful selection of all hardware and software items that you wish to maintain control over. "Maintaining control" means that at any given point in time you are able to report on what the system currently consists of, what changes are being proposed, what changes

**TABLE 8-1 Configuration Management Elements—Identification**

| | |
|---|---|
| Definition: | Careful definition of baselines and baseline components for both HWCIs and SWCIs to be managed |
| Question: | What is my system configuration? |
| Answer: | Your system configuration consists of the following items—(Item 1, Item 2,...) |

**TABLE 8-2 Configuration Management Elements—Configuration Control**

| | |
|---|---|
| Definition: | The mechanism for preparing, evaluating, accepting, or rejecting proposed changes |
| Question: | How do I control changes to my configuration? |
| Answer: | The steps in processing changes in my configuration are: (Step 1, Step 2,...) |

**TABLE 8-3 Configuration Management Elements—Auditing**

| | |
|---|---|
| Definition: | The mechanism for determining the current state of your system |
| Question: | Does the system I am building satisfy the stated needs? |
| Answer: | Your current system differs from the stated needs as follows: (Difference 1, D 2,...) |

**TABLE 8-4 Configuration Management Elements—Status Accounting**

| | |
|---|---|
| Definition: | Tracking and reporting all identified HWCIs and SWCIs and other identified items |
| Question: | What changes have I made to my system? |
| Answer: | Your configuration and changes to your system at this time are: (Item 1, Item 2 ... and Change 1, Change 2, ...) |

have been accomplished, and what effects, if any, these changes have had on the ability of the system to meet original requirements.

The process also involves defining baselines appropriate to your work and defining the extent to which CM tools will be applied at each baseline.

***Identifying Baselines***

Baselines are established at discrete points throughout the implementation process to provide increasing degrees of control as the system product matures. As this maturation unfolds and further specific commitments are made, it is both possible and desirable to increase your control over the emerging levels of detail.

There are at least five logical discrete points that can be used to establish baselines for the purposes of CM. You may elect to use all of them or a subset, depending upon the size, complexity, and length of your project and upon your level of confidence that all facets of the work can be adequately tracked. They are the functional, allocated, design, product, and operational baselines, as shown in Figure 8-1. The following paragraphs describe these baselines.

Functional baseline—The functional baseline is typically established at the conclusion of the formal establishment of system functional requirements. The Functional Requirements Document is established prior to the system requirements review and in many cases provides the basis for a request for proposal.

Allocated baseline—The allocated baseline is normally established after the preliminary design review and prior to initiation of detailed design. It represents the point in the engineering process where required system performance has been allocated through system specifications to specific hardware and software configuration items and subsystems.

Design baseline—The design baseline, when used, is established following detailed design and at the conclusion of the critical design review. This is

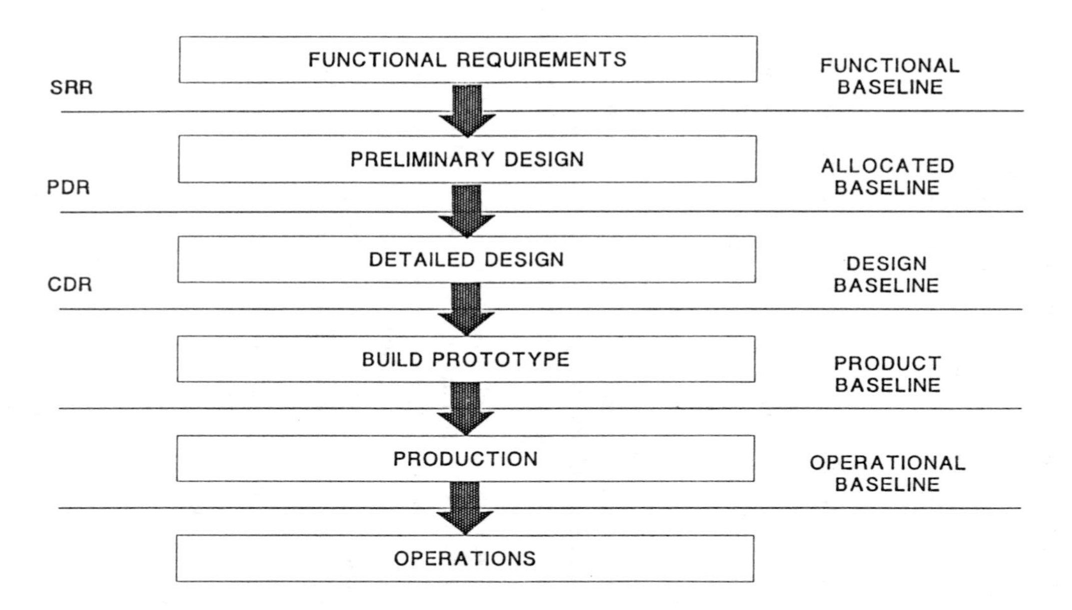


**FIGURE 8-1.** System life cycle baselines.

the baseline to which all subsequent hardware and software implementation adheres.

Product baseline—The product baseline is established prior to production. System and subsystem specifications, initial fabrication, system integration, and test and factory readiness are all complete.

Operational baseline—The operational baseline is established as the system is delivered to the users and is the basis for all continuing system maintenance and product improvement phases.

Military programs typically use the functional, allocated, and product baselines. The functional baseline is established at the completion of the *conceptual exploration* phase. The allocated baseline is established after the *validation* phase and early in the *full-scale development* phase. Product baseline hardware is established prior to production, and product baseline software is established at the completion of software code and test.

As each baseline is established, control of changes to that baseline is also established. Formal methods of change control are discussed below under "Configuration Control."

***Identifying Configuration Items***

The items selected for control under configuration management are called configuration items. Logically, there are both hardware configuration items (HWCIs) and software configuration items (SWCIs). In the Department of Defense system hierarchy, these are usually defined at level 4 of the system at the hardware subsystem and software program levels, as depicted in Figure 8-2. This is a very logical place to start in nonmilitary systems as well in an effort to identify that which you intend to maintain control over.

However, as always, the selection is not blindly driven. Rather, it is driven by the level of confidence in the degree of control required. The systems engineer may wish to maintain close control over critical subassemblies or program components. Alternatively, control at the higher segment or element levels may be perfectly appropriate if these items are off the shelf and proven and if confidence is high that little change will take place.

The identification of configuration items is very important and worthy of considerable thought. Any given system may not conveniently map into or onto a traditional system breakdown found in DOD or other standards. The work breakdown structure, at all levels, can also be used and is an effective guide in helping to consider and identify all items. The judgement of the systems engineer here is key. Configuration items should be identified such that adequate control will result but the total number of items will still be manageable. Also consider such aspects as level of functionality, testability, boundaries of discipline, and division of responsibility to distant organiza-

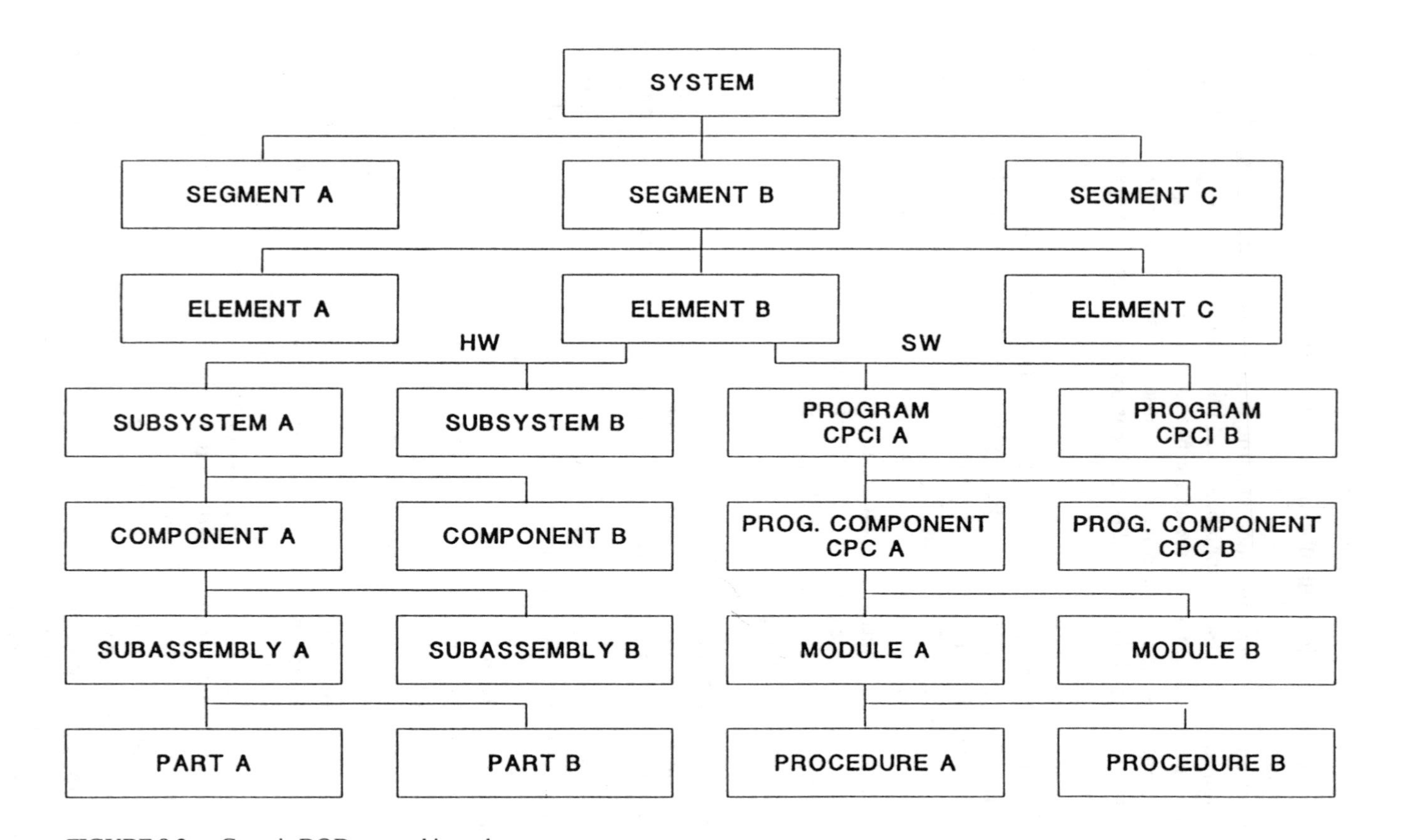


**FIGURE 8-2.** Generic DOD system hierarchy.

tions or contractors. Note that these considerations are similar to those evaluated in the allocation of work to specific subsystems. As always, the level of confidence with regard to maintaining visibility should assist in determining configuration item (CI) selections. It is appropriate to review CI selection with the SDT and to seek the team's concurrence.

Also recognize that many organizations today routinely employ CM strategies designed to support the staircase with feedback model. If the selection of baselines and CIs is predetermined by outside influences and it is felt that good reason exists to deviate or seek relief from an established policy, the systems engineer should carefully consider alternatives and propose them to both CM management and line management rather than knowingly commit to definitions and procedures that may hamper the ability to perform.

## CONFIGURATION CONTROL

Configuration control encompasses the formal set of policies and procedures for requesting, evaluating, and accepting or rejecting proposed changes, corrections, and/or waivers to a configuration item.

### *Change Control*

Figure 8-3 displays the salient steps in the generic process of change control. Requests for change can be driven by any unforeseen need, such as necessity for design changes, response to changing funding profiles, schedule changes, and other *engineering change proposals* (ECPs). Requests for change can also be planned, such as installation of new operating systems, placement of new software deliveries on line, and implementation of preplanned product improvements.

Configuration control is maintained on the current baseline. The current baseline includes itself and previous baselines so that ECPs may affect any or all of the functional, allocated, design, product, or operational baselines, depending on the time they are requested. Thus, changes that affect physical hardware and software also result in changes to all supporting documentation as well.

A single responsible person is designated to log and track each ECP. This person can be the systems engineer, the design team secretary or any other member of the design team. On large projects, the responsibility may be assigned to a representative of the organizational CM structure.

Assuming the submission is a valid change request and not simply a failure report or a misunderstanding, the design team then evaluates the impact of the ECP. The design team usually restricts itself to evaluating technical impacts. Potential programmatic impacts, such as schedule and cost impacts,

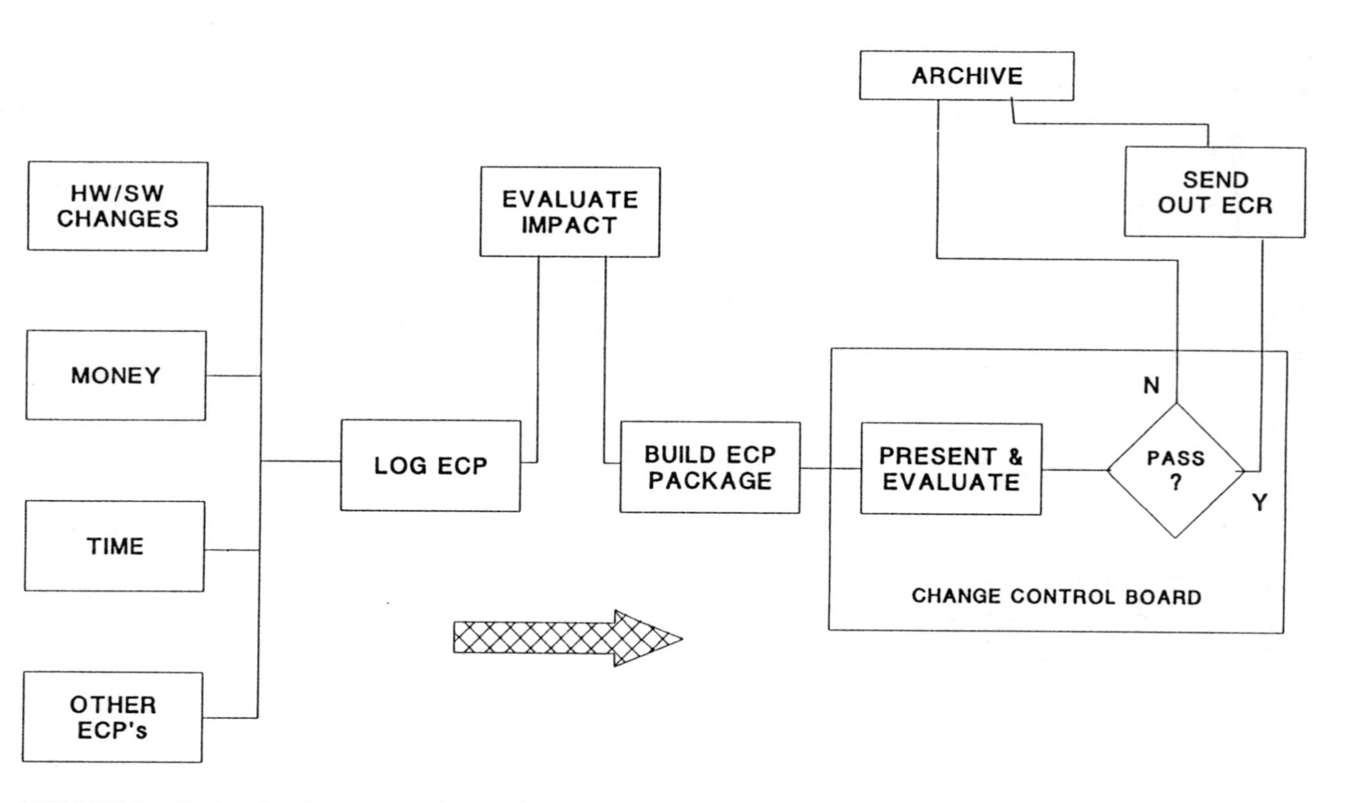


**FIGURE 8-3.** Engineering change proposal processing.

should also be brought to the attention of project management, and the design team then supports management in assessing the overall impacts.

The design team then prepares or supports the preparation of an ECP package that will be presented to the formal change control board (CCB). CCBs usually have a number of members not directly connected with the project, which is both good and bad. It is good because, at least in theory, an element of objectivity is introduced into the deliberations. It is sometimes bad because the necessary understanding to make or break the case is not always present. While this may at times be frustrating, the condition is largely offset by requiring the proponent to make his or her case in a thorough manner. The best CCBs have representation from your CM organization, your project and programmatic management, the system user, all affected technical areas, and a technically competent "outside" expert.

The system engineer often makes the presentation to the CCB, although the system user may also be the presenter. Take the appropriate COGEs with you, as well as any other specialty technical backup you may need to respond to detailed questions.

Your organization may have a formal form and format for CCB presentations. In any case, be sure your presentation includes:

1. A statement of the problem or system upgrade proposed and a statement of the change requested. The latter should include the exact choice of language for the proposed change(s) to the affected specifications and/or requirements, if appropriate.
2. A statement summarizing the analysis of all impacts, including technical, programmatic, and any user impacts. Don't forget the latter—it is the most important.
3. Tabulate on a separate slide the advantages and disadvantages of ECP implementation versus alternatives, even if an entry under one of the headings is none.
4. Your recommendation to accept or reject the ECP, supported by a succinct statement of consequences.

Don't give a lengthy presentation. Go through the above items logically and sit down. Save the backup details for responses to specific questions from the board. A basic presentation may consist of 4 to 6 slides, with up to 20 backup slides.

Accepted changes are often classed with respect to their impacts and priorities. The priority largely determines the time frame in which the change will be implemented. Two common priority schemes for handling of ECPs and FRs are listed in Table 8-5. The classifications generally effect the importance and hence the timing of the change to be made.

**TABLE 8-5 Common Priorities for Handling Change**

| |
|---|
| **Two-level** |
| 1. A change is required in form, fit, or function. |
| 2. Changes are required in document language or in minor component substitutions. |
| **Four-level** |
| 1. Prevents mission accomplishment. |
| 2. Seriously degrades mission accomplishment. |
| 3. Minor impact on mission accomplishment. |
| 4. Documentation change only. |

Following the CCB meeting, an engineering change request (ECR) is issued by the board if the proposed change has been accepted. The ECR constitutes official authorization to generate detailed change implementation plans and to initiate the change. The progress of change implementation is tracked by a single, designated responsible person. All ECPs and ECRs are archived by CM.

CCB meeting minutes are also recorded by a designated CCB secretary. The minutes should include a list of all in attendance, their affiliations, a brief but complete account of proceedings, and any technical or programmatic supporting material necessary to provide a self-contained representation of what transpired.

### *Failure Reports*

Failure reports (FRs) are also logically handled by CM for at least two reasons. First, a record of outstanding FRs and closed FRs is a widely used metric for partial reporting on the status of an implementation effort. Secondly, it is common for FRs to be written, with request features that are beyond the stated requirements—that is, requests that actually should have been change proposals.

In principle, FRs can be submitted by anyone. The point in time at which a "failure" occurs, however, is usually considered to be *after* formal delivery of hardware or software. The term "delivery" refers to a specific point in time at which developers consider the product to be complete and hand it over to another party, such as an independent testing group, or to end users for acceptance testing. That is, programmers in one development group typically don't generate FRs upon themselves. Their interactions are generally less formal—as they should be, to allow responsiveness in identification and correction of bugs.

In more rigorous settings, FRs can be generated during development by a programmer in one organization against a programmer in another organization. This can be desirable when large programming staffs are involved across organizational entities as specific programs are provided for cross support. This practice can also be desirable for keeping a record of problems encountered when building novel systems for the first time.

When programmers interact more intimately, however, there is a fine line between simple debugging in unit test and identifying a "failure." A failure is quite naturally not perceived as a failure during debugging activities.

An alternative approach to record keeping during development, purposefully designed to be less stringent, involves the use of anomaly reports (ARs). ARs are used primarily to maintain a log of problems encountered such that the experience can be put to good use in the future. This is a good idea in theory, but is difficult to implement effectively. Programmers don't like to fill out forms. Still, it is often useful to maintain records of significant unforeseen difficulties that may help you and your programmers in future similar efforts. The use of ARs is one method of doing this. Another method, which we all should have learned in school, is to have your programmers maintain the classic engineering notebook. This is a good habit and is particularly valuable when dealing with new designs—that is, designs that involve the organization of knowledge, algorithms, or data structures not encountered before.

A standard method for handling failure reports is depicted in Figure 8-4.

FRs, as they are generated, are routinely submitted first to the SDT, where they are logged. On small projects, this may be done by the systems engineer. On larger projects, designate the design team secretary to log and track all FRs.

An initial assessment of the FR is made to determine:

1. Whether the report records an actual failure;
2. Whether it requests a change to existing requirements; or
3. Whether the FR was generated by a misunderstanding of how the system properly operates or should be used.

By definition, a failure is specifically a failure to meet a single, measurable, documented functional requirement or specification derived therefrom.

There is a fourth reason for rejecting an FR. The FR should address a single issue. An FR form that addresses multiple problems by listing more than one at a time should be resubmitted, such that there is a single form for each perceived failure. The reason for this is that, as each problem is successfully addressed, the identified FR can be closed and partial closures of FRs can be avoided.

The systems engineer is responsible for each of these four determinations. The status of each submitted request should be routinely reviewed with the SDT.

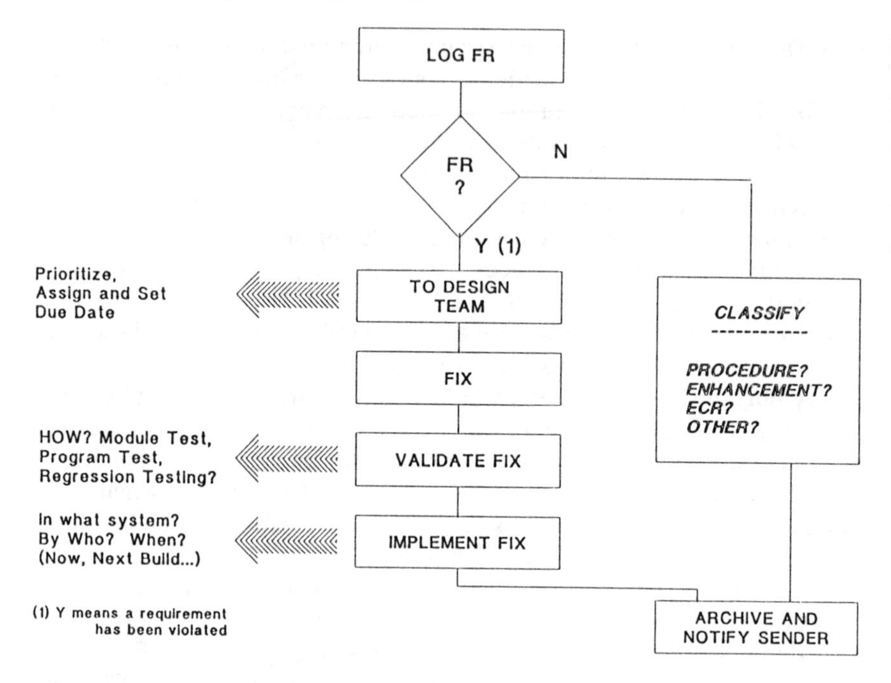


**FIGURE 8-4.** Failure report handling.

If the submission is not a valid FR for any of the four reasons given above, the reason is noted on the appropriate section of the FR form and the originator receives a copy. The generator may then take appropriate action in response. All FRs, regardless of their content, are archived.

If the FR calls out a single bonafide failure, a plan of action is then agreed upon by the design team. This action consists of setting a priority for correction, designating an assigned party responsible for correction, and setting a correction due date.

From this point on, there is only one copy of the FR that is official. The systems engineer or a designee acts as sole custodian for the official version. Any copies of the official copy that are made during the interim course of activities prior to FR closure are not official copies. Copies are distributed for information purposes only, with regard to the latest status of progress toward closure. Entries in fields of the FR form made on distributed copies have no validity. Official entries of information in any field of an FR form can only be made by the systems engineer or by a party designated by the systems engineer, and only on the original. If a separate organizational CM entity is tracking your FRs, you must work out an agreement for a strategy of control for a single official FR.

On a given project, priorities for correcting valid FRs are generally the same as those adapted for handling CRs, as shown in Table 8-5.

The responsibility for correction is generally assigned to the COGE for the area affected. This does not mean that the assignee will necessarily make the correction firsthand. He or she may assign the task to someone working for them. But the assignee is solely responsible for the correction being done and tested at the unit level on or before the agreed-to due date. The person making the correction is not responsible, unless that person is in fact the assignee.

The setting of the due date is primarily driven by the needs of the user and is negotiated, as are priorities and correction responsibilities, at the SDT meetings. Potential schedule impacts on other concurrent work need to be identified and discussed. The guiding concept in all of these determinations should be to accommodate the user as much as possible. But also be very sure that the assignee is comfortable with the due date. You must communicate on this point. Don't force an unrealistic achievement upon the assignee. Bean counters love late FRs, they absolutely love them—and they will find you.

While the fix is being carried out, a plan for acceptance testing of the fix and its ultimate implementation on the system must be specified. These plans can be quite simple, or they can be genuinely complicated.

On projects of any size, testing and implementation schedules are usually determined by previously scheduled build deliveries.

Realistically, cases sometimes arise when even our best priority schemes become replaced with "do it immediately" and deviation from accepted practices suddenly becomes your order. In the "quick fix" setting, the trick is to determine whether the fix is related to a small and isolated hardware or software condition or whether the fix has system-wide effects. The great danger is assuming the former when the latter is the case. The safe approach, as always, is to do a lot of testing at the development level—right at home—before implementation in the operations environment. If a quick fix is truly required, assemble the SDT together for a special meeting. Discuss all aspects of the problem. Coordinate the special actions you must take by making specific assignments, and devise contingency support plans should the issue persist. Ask for and use all the time you can get.

### *Liens*

Liens consist of promised hardware items or software capabilities that remain outstanding at the time of a delivery. They are usually fairly minor items for which workarounds have been devised. Liens are accepted (tolerated) by entities that are receiving hardware or software deliveries because it may be more desirable to proceed with scheduled testing or operations without the full complement of system capabilities than to wait for all promised items to be provided. Thus, the decision to execute a delivery that is deficient in one or more items is primarily made by the entity that is to receive the delivery.

Liens arise for a number of reasons. Procurements may be late due to vendors slipping deliveries to your project, financial decisions out of your control, or any number of procurement problems. If previous deliveries have already been made, there may be outstanding failure reports that have not been corrected at the time of delivery. Liens in one form or another against documentation errors or deficiencies are also common. Action items generated at any of the reviews held prior to the first delivery or between deliveries may not have been adequately resolved. Finally, unforeseen technical difficulties can easily cause schedule changes that impact the ability to deliver all capabilities on schedule.

Note that liens are written against requirements. They are a part of an agreed-upon strategy to forgo having to meet a requirement at the time of a delivery. Thus, change requests cannot become liens unless the change has been formally incorporated by the configuration management change control board as a new requirement to be met by the project and a delivery date is to be missed.

Action to remove liens is negotiated by the SDT in the same manner as failure reports are. Each lien is assigned to a responsible party for its resolution, a priority is assigned, and a due date is established. For consistency, it is expedient to use the same priority scheme that you use for failure reports, change requests, and action items.

A listing of liens is maintained by the systems engineer. An example of a lien list is provided in Figure 8-5.

The sample lien list shows outstanding liens as of 6/23/92. As each lien is closed, its closure is noted in the next reporting period and is then subsequently dropped from the list to avoid carrying excessive outdated information. The complete history of lien handling, however, is maintained in a computerized lien list data base. The reporting periods are generally 1 week apart, as the lien list is routinely reviewed at each design team meeting.

An identification number for each lien is given in the first column. The second lien in the list, lien number 5, was closed during the week prior to 6/23/92. Implicit in the list is the fact that liens 2 through 4 and lien number 6 were closed in earlier reporting periods and that these records were dropped from the data base report form. In the next reporting period, lien number 5 will be dropped from the list.

Lien number 8 is currently past due. The reason for this condition is summarized in the comments column, and a new anticipated closure date is shown. Lien number 9 calls for a documentation section rewrite for clarifi-

Page 1 of 1
Report Date: 6/23/92

PROJECT ABC LIEN LIST

| No. | H-Hardware S-Software | Assign Date | Lien | Assigned To | Due Date | Comments |
|---|---|---|---|---|---|---|
| 1 | S | 2/4/92 | PACK routine | J. Smith | 7/12/92 | |
| 5 | S | 2/12/92 | COPY routine | R. Jones | 6/19/92 | Closed |
| 7 | S | 3/18/92 | Tape to Tape routine | M. Gorden | 8/13/92 | |
| 8 | H | 5/2/92 | Terminal | L. Scott | 6/1/92 | Vendor delay, six of seven delivered, seventh due 7/15/92 |
| 9 | S | 5/2/92 | Users Guide Document | J. Smith | 7/1/92 | Clarify Sec. 7.1, See AI #23 |
| 10 | H | 5/15/92 | Range Bin #4 | L. Scott | 7/15/92 | In repair. |

**FIGURE 8-5.** Lien list example.

cation that resulted from a review Action Item (AI #23), and lien number 10 resulted from an unforeseen hardware failure.

## AUDITING

Configuration auditing is the process of determining the extent to which the current configuration baseline conforms to the previous baseline and to the original functional requirements (functional baseline). The process involves both *verification* and *validation,* so called V and V. These terms are widely used, but often interchanged and more often misunderstood.

Verification is the process of determining if the configuration items of a given baseline meet the requirements of the previous baseline. This is needed, but note that simply verifying that the more developed CIs of a given baseline are traceable to a previous, more generic baseline does not guarantee in itself that established baselines have not strayed through time from the original intent of the system. That's where validation comes in.

Validation is the process of determining if a given baseline still solves the intended problem. Validation is concerned with traceability to original mission objectives or, putting it more simply, does it work right?

Both of these functions are best tracked by using traceability tables. For the purposes of verification, the tables trace forward and backward between baselines. The validation tables trace each baseline, as it is established, both forward and backward to the original functional baseline.

Two sample tables to assist in validation are given in Tables 8-6 and 8-7. Table 8-6 shows a portion of the first table, which consists of a succinct summary of functional requirements. This table is constructed directly from the functional requirements document and serves as an easily handled quick reference for each requirement. The functional requirements document itself, of course, contains each requirement in its entirety and can be referenced as needed for clarity.

The requirements summary table provides a terse description of a single requirement for each appropriate paragraph in the FR.

The traceability table for each baseline accompanies the requirements summary table and records the specific action taken at a given baseline in response to each original requirement by paragraph. An example of a portion of a validation traceability table for a typical design baseline is given in Table 8-7. The example shows traceability of each FR paragraph to a paragraph in the *software specification document* (SSD) as well as the hardware items and software modules intended to meet the requirement. Thus, the software program called QPLOT1 is shown to respond to the functional requirement for Generation of Plots. Note that the SSD paragraph numbers are not necessarily in sequence, as are the FRD paragraph numbers. This point

**TABLE 8-6 Sample Portion of Functional Requirements Summary**

| FRD Paragraph | Requirement |
|---|---|
| 1.0 —— | |
| 2.0 —— | |
| 3.0 —— | |
| 3.1 | APPLICATIONS PROCESSING REQUIREMENTS |
| 3.1.1 | Generation of Plots |
| 3.1.2 | Data Conversion from ASCII to EBDIC |
| 3.1.3 | Contrast Manipulation |
| 3.1.4 | Two-dimensional Fourier Transforms |
| 3.1.5 | Map Projections |
| 3.1.6 | Photographic Hardcopy |
| 3.1.7 | —— |
| 3.1.8 | —— |
| 4.0 —— | |
| 5.0 —— | |

**TABLE 8-7 Sample Portion of Design Baseline Traceability**

| FRD Paragraph | SSD Paragraph | Hardware | Software | COMMENTS |
|---|---|---|---|---|
| 1.0 —— | | | | |
| 2.0 —— | | | | |
| 3.0 —— | | | | |
| 3.1 | | | | |
| 3.1.1 | 4.2.3 | | QPLOT1 | |
| 3.1.2 | 5.1.1 | | AECONV | |
| 3.1.3 | 5.2.4 | | CONMAN | |
| 3.1.4 | 5.2.5 | | FFT101 | |
| 3.1.5 | 6.3.1 | | MAPPR2 | |
| 3.1.6 | 4.3.3 | FOTO | FOTOHC | Two FOTO units to be used. |
| 3.1.7 | —— | | | |
| 3.1.8 | —— | | | |
| 4.0—— | | | | |
| 5.0—— | | | | |

highlights the fact that the traceability carried out in this example is forward, from the FRD to the design baseline.

Backward traceability consists of making sure that all baselined hardware and software items have been accounted for by forward tracing and that no items exist for which there is no documented need. This exercise can be carried out using traceability tables of similar construction but that are designed to trace in the opposite direction. Both forward and backward tracing should be accomplished in the interest of completeness.

The three right-hand columns of the sample traceability table indicate the specific hardware items and software modules that respond to each requirement and also provide an area for comments.

Such tables are intended to assist in organizing one's thinking in carrying out the auditing process. The mechanical use of these tables does not, in itself, constitute auditing. Careful thought regarding the adequacy of elements as they represent an increasing level of detail in successive baselines is the real function of auditing. Successive levels of detail should be clear and unambiguous, such that a lucid forward and backward mapping between baselined items and to and from functional requirements is evident.

Effective auditing insures:

1. That the unfolding design and implementation is responsive to actual needs; and
2. That unspecified capabilities are not finding their way into the system.

## Configuration Management of Vendors and Subcontractors

Large vendors usually have their own CM organizations and practices. Small ones may not. When issuing a request for proposal or a request for quotation, it is advisable to include a separate section asking prospective vendors to state how they intend to meet your project's CM requirements. State generic CM requirements clearly but not in such detail as to require major modifications in the way the vendors currently do business. Evaluate vendor responses numerically as a separate part of your vendor acceptance criteria algorithm. If all aspects of a vendor's response are satisfactory except their approach to CM, you can always impose your own organization's system on them. It is wise to avoid this, however, as it is only likely to increase the cost.

I can recall observing a procurement in which an elaborate aerospace CM system was imposed upon vendors in an RFP for an earth-based system. The prospective vendors were all large, capable corporations, but the CM practices they were being asked to follow did not map at all into their existing capabilities. The unanticipated result was that every bid received was prohibitively high in cost—so high that the RFP had to be recalled and reissued. It was subsequently discovered that the major contributor to the high cost was the perceived extra effort required to meet the specific CM requirements of the procuring organization, which detailed procedures down to report types, forms, formats, and specific processes.

Good corporations know how to do CM in their fields. Contractors and vendors should be given an opportunity to specify how they intend to meet the project's stated CM needs. Negotiate, if required, with the idea of maintaining their basic approach and procedures. Finally, include in the selection criteria a means of eliminating those who are unable to maintain effective configuration management.

## The Configuration Management Plan

The configuration management plan (CMP) formally defines and describes the policies and procedures to be employed in the execution of CM. The CMP also defines organizational structures and responsibilities for CM functions. This section provides a guideline for constructing the CMP.

The CMP may be imposed on a project by an external CM structure by directive. If this is the case, it should be perused with great care. CMPs written to apply to any and all projects can be designed to cover all the bases in such detail as to result in an overbearing regimen. Still, such documents are valuable as guides. Should the systems engineer feel that specific modifications or waivers are in order and can present sound reasons for their adaptation, it is not unreasonable to request such actions. Good CM plans balance the need on a project-by-project basis for individual (or team) creativity against the need for increasing levels of control through time.

The more creative phases of a project occur earlier in the development cycle. The need for creativity diminishes as the design matures. The most freedom is required between the development of functional requirements and the preliminary design. While strict control is maintained over the functional requirements document, the freedom to determine issues of "how" in response to the "what" of the functional requirements should ideally be maximized within the functional requirements envelope. In the staircase with feedback systems engineering model, the work accomplished between the preliminary design and the detailed design is naturally restrained to a more confined envelope. There are less options, and the need for control becomes increased. Beyond the detailed design after passage of the critical design review, control typically needs to become rather stringent. This is a simple top-level concept and should be a consistent element in guiding the approach to CM implementation.

Table 8-8 provides a suggested outline for a generic CMP.

The following paragraphs amplify the content of each of the above sections.

**TABLE 8-8 Configuration Management Plan Outline**

1. Introduction
2. Organization/responsibilities
3. Configuration identification
4. Configuration control
5. Configuration status accounting
6. Configuration auditing
7. Subcontractor/vendor control
8. Glossary of terms

Introduction—Contains a one-sentence statement of the document's purpose. Include a statement of scope, if appropriate, and include a subsection covering applicable documents.

Organization/responsibilities—Show the organizational relationship of CM to project management, and show an organizational chart of the CM function. Include the change control board's organizational relationship. Discuss responsibilities of management, the CM organization, and your design team.

Configuration identification—Define the baselines to be used in the project. Clearly define the beginning and end of each baseline in terms of deliverables, completion of documentation, and/or completion of reviews. Explain the philosophy to be used in designating hardware, software, and documentation as configuration items. (The actual designation is the execution of the plan.) Specify the plan for use of identifying numbers for each level of configuration items.

Configuration control—Explain the plan for the systematic evaluation, coordination, approval/disapproval, and subsequent handling of proposed changes to any baseline. Define the entire change review and implementation process as discussed in the section above, tailored or expanded to your need with a supporting block flow diagram. Cover all functions of the change control board, including the CCB review of deliveries and placement of hardware and software on line. Provide sample forms in an appendix for each form you intend to use, such as failure reports, requests for change, anomaly reports, documentation change notices, implementation notices, and so on. Include directions on the reverse side of each form as to how to fill them out, the purpose of the form, and the organizational entity to submit them to.

Configuration status accounting—Specify the form and format for each item covered in Figure 8-2 as a minimum. Specify all data base characteristics, such as sorting capability and report generation capabilities, that you may require in order to conduct full or partial audits.

Configuration auditing—Define all schedulable formal audits. As a minimum, a formal audit should be scheduled in support of the acceptance of each baseline. Specify the need for unscheduled partial audits on demand that you may wish to conduct, such as statistics on FR closures, changes in progress, and changes implemented.

Subcontractor/vendor control—Specify your plan to control subcontractors and vendors. If you are using their plans, provide a summary description and refer to their documentation. Describe modifications to your own control procedures that are acceptable and have been agreed upon. Describe all interfaces and contacts between your CM status accounting process and their's. Describe all auditing practices, scheduled and unscheduled, that you intend to carry out on vendors and subcontractors.

Glossary of terms—Do not assume that terms such as lien, failure report, baseline, and so on are well understood. Include a glossary of all pertinent terms.

## STATUS ACCOUNTING

As changes to the system occur over time, you will begin to amass pieces of paper, such as ARs, FRs, CRs, and other formal requests for correction and change. It is not sufficient to simply collect these in folders, corners, and drawers, such that constant shuffling is required to determine the latest status

**TABLE 8-9 Content of the Status Accounting Data Base**

1. A description of each CI.
2. The location, identification number (if appropriate) and COGE for each CI.
3. A listing of each CI in each baseline and the date that each baseline came into being.
4. ECRs and their status as of the reporting date. ECR status includes approved, disapproved, progress of approved change implementation, responsible party, priority, and due date.
5. FRs and their status as of the reporting date. FR status includes responsible party, priority, and due date or closure date. These records also provide a base for tabulating FR generation and closure rates.
6. Action items generated as a result of audits or formal reviews and their status.
7. Any other information you may deem advisable as a result of configuration identification, configuration management, or auditing functions.

of configuration items. The systems engineer must be capable, on demand, of easily updating configuration status and generating a succinct report for his or her own use and for management's use.

Status accounting is an administrative function that maintains a formal record of each of your configuration items, including their definitions, change requests, change approvals and disapprovals, progress toward implementation of approved changes, implementation completions, and actions taken as a result of configuration audits.

Your status accounting mechanism should be computer based to support the organization of what can become large amounts of data and to facilitate auditing and reporting functions. A computer-based system also allows you as the systems engineer to stay current in a convenient manner.

The status accounting data base should include, as a minimum, the data shown in Table 8-9.

# 9

# Interface Definition and Documentation

A fundamental challenge to the systems engineer is to maintain control over the definition, development, design, and implementation of system and subsystem interfaces. There may be domains within the realm of systems engineering where discussion occurs regarding the responsibility of the systems engineer. However, there is virtually universal agreement that system and subsystem interfaces are a major responsibility of the systems engineering function.

The point is emphasized through the soccer ball analogy shown in Figure 9-1. The outline of the ball represents the system boundary, which receives inputs from and delivers outputs to the outside world. The seams in the soccer ball depict the boundaries between subsystems within the system. The systems engineer is responsible for defining and controlling the flow across both system and subsystem boundaries. These are the lines upon which the systems engineer must walk with significant authority as well as responsibility.

While the subsystem COGEs are deeply imbedded in the details of subsystem design and fabrication, the systems engineer typically follows this detail only to the extent necessary to insure that the integrity of the interfaces is maintained. This detail, however, must be pursued with vigor.

This chapter presents an organized methodology for developing of interface definitions, and provides guidelines for interface documentation.

## THE $N^2$ DIAGRAM

The $N^2$ diagram represents a widely used structured method for the initial definition of interfaces as well as their successive detailed definition at lower levels [1]. The general structure of the $N^2$ diagram is shown in Figure 9-2.

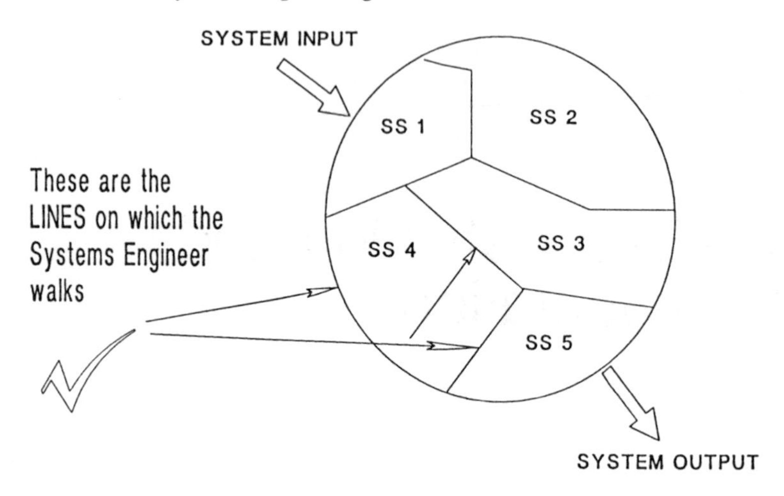


**FIGURE 9-1.** Think soccer ball.

Subsystems are positioned on the diagonal of the diagram. Subsystems A through F are depicted in the figure. Subsystem outputs are displayed as horizontal lines, and subsystem inputs are displayed as vertical lines. The absence of a dotted box at the appropriate location of the matrix means that there is no direct interface between the designated subsystems.

In the example of Figure 9-2, subsystem A has an output that becomes an input to subsystem C, denoted by the dotted box with text "from A to C." In fact, subsystems A and B both have inputs to and outputs from subsystem C, but note that subsystem A has no direct inputs to or direct outputs from subsystem B. Similarly, subsystem D has inputs to subsystem C, but has no direct inputs from subsystem C. Subsystem F has outputs that become inputs of all other subsystems.

The interface definition process begins at the top level and proceeds to lower levels. Iteration between levels takes place as more insight is gained and design decisions are made. Initially, the interfaces are described in English, with more detail added at each successive level. Interface particulars are eventually expressed in precise detail, using tools such as program design languages, circuit data sheets, or engineering drawings.

By way of example, consider the Automated Meteorological Data Acquisition and Reporting System (AMDARS) discussed in Chapter 5, "The Work Breakdown Structure." Recall that the function of the AMDARS was to automatically gather specific weather data at remote marine sites and periodically transfer current data readings to a central site.

The top-level AMDARS $N^2$ diagram is shown in Figure 9-3. The system is composed of a *central segment* and a *remote segment*. The central segment

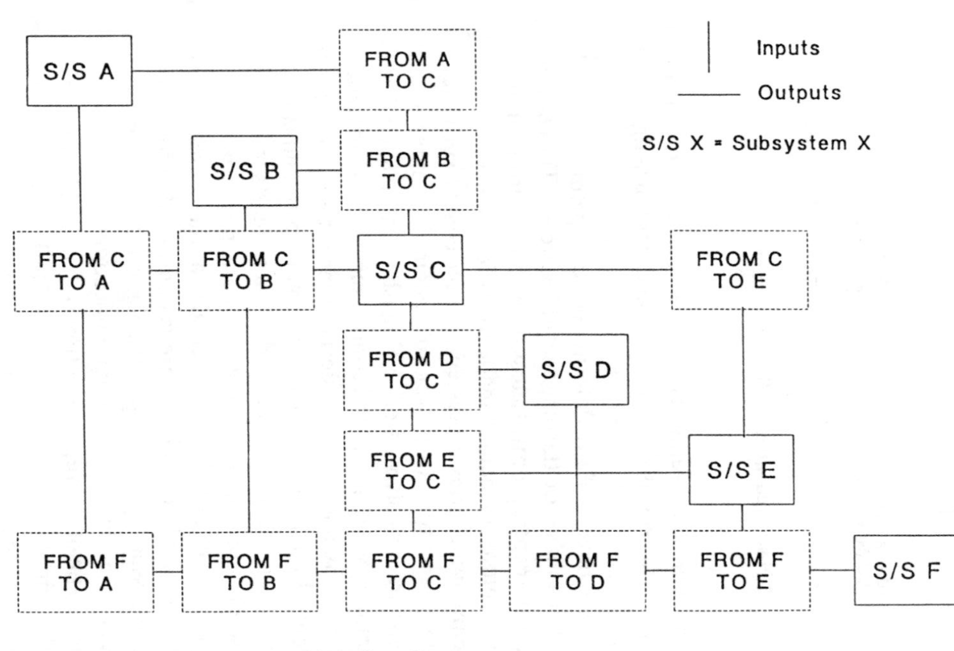


**FIGURE 9-2.** The N-squared interface diagram.

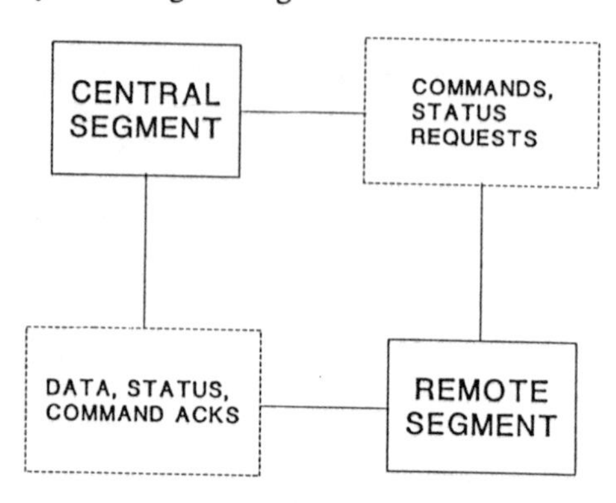


**FIGURE 9-3.** Level 1 N-squared interface diagram example.

outputs command and status requests to the remote segment, and the remote segment returns data, status information, and acknowledgements to the central segment.

Assume that we have initially determined to implement the AMDARS with the subsystems shown on the diagonal of Figure 9-4. The central segment is to be made up of a central control subsystem and a central communications subsystem. The remote segment is made up of four subsystems, designated as the remote communications subsystem, remote control subsystem, remote sensor subsystem, and remote power subsystem. The interfaces between these subsystems are initially written in English for design team review.

In this preliminary view, the central control subsystem sends data requests, sleep/wake commands, and status requests to the remote control subsystem through the central communications and remote communications subsystems. The remote control subsystem sends data requests to the sensor suite in the remote sensor subsystem and receives data in return.

It is useful to observe certain characteristics of the exercise so far. While it has been a fairly simple exercise, there is considerable focus on the next level of significant issues. For example, the remote sensor subsystem has no direct interfaces with the remote communications subsystem. A further focusing characteristic of the representation is that the $N^2$ structure immediately enables the remote control subsystem COGE to identify the need for negotiation between the communications COGE and the remote sensor COGE, to further define information data formats. One obvious consideration is where to digitize sensor data. If it takes place in the remote control subsystem, the remote control COGE will benefit from further definition of that subsystem's role. Another highlighted issue relates to how status data is to be assembled—that is, shall built-in test equipment (BITE) be used, or can sufficient status be gathered for modular replacement purposes from the

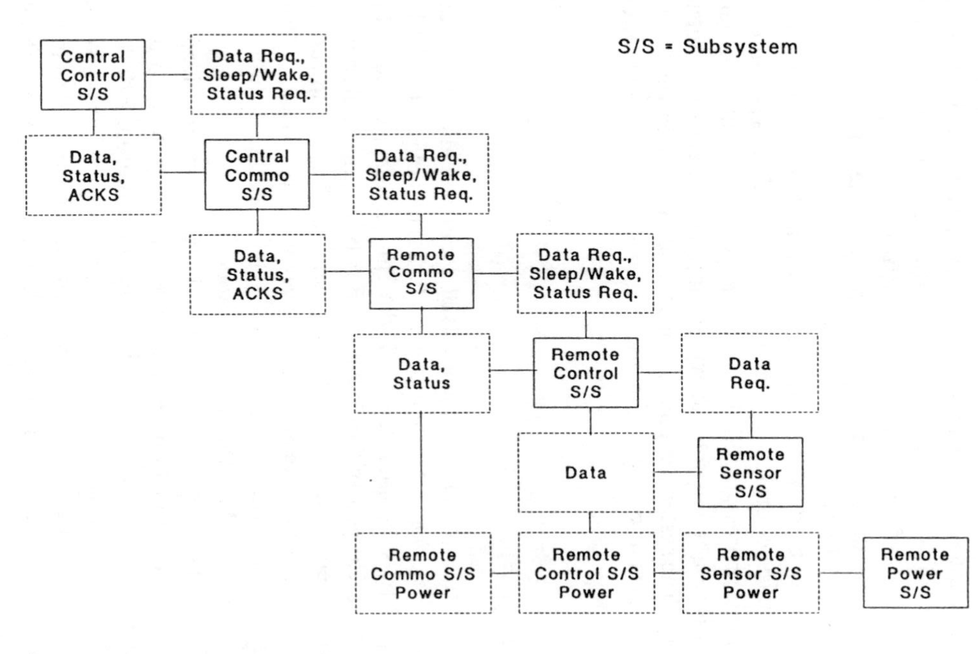


**FIGURE 9-4.** AMDARS interface diagram example.

logical interpretation of information flow between subsystems (readings out of range, lack of communications acknowledgements, etc.). It is also clear that communication protocols can now be addressed between the specific subsystems affected by them.

Another important characteristic that should be evident at this level of discussion is that the interface descriptions remain generic to these more detailed design issues. What should be happening is that a clear delineation of the next level of design issues logically surfaces as a direct result of the $N^2$ formulation. The formulation represents a structured methodology—hence an opportunity for the systems engineer to direct the unfolding of needed detail in a logical fashion.

The process is carried forward with more and more detail involving negotiation and issue formulation by the COGEs and the systems engineer. Throughout this process, the SDT is the focal point for identifying pertinent issues and assigning action items to explore approach alternatives outside the SDT meetings. The systems engineer leads this identification and assignment effort, with particular care given to addressing all issues at the proper level of definition.

## THE OSI MODEL

One useful means of visualizing the levels of detail in dealing with interfaces is provided by the *open system interconnect* (OSI) model [2]. The OSI model was developed by subcommittee 16 of the International Organization for Standardization. The goal of the OSI model is to bring universal compatibility between computers, computerized equipment, and communications networks, from vendor to vendor and ultimately from nation to nation. The term "open" refers to the concept that any system developed with the OSI model would be open to all others that used the model.

The OSI model consists of seven layers. These are summarized in Table 9-1. Examples of the functions at each layer are given to help determine the applicability of each layer to a given implementation. Efforts are still underway to bring precise definitions to the upper levels. While this resolution is still in progress, the user can generally be viewed as interacting at the application level when all seven layers are used. Some systems require a lesser number of levels. For example, a computer-to-computer dedicated communications system may require the bottom three levels only, involving specification of the electrical connection, a suitable protocol, simple addressing, and error control. In this case, the user would view the ISO model at the network level. On the other hand, a telephone system involves simply the physical and network levels. That is, there is no

**TABLE 9-1 Open System Interconnect Model Layers**

| Layer | Function | Examples |
|---|---|---|
| Application | Direct support of user, system management | Batch, data entry, DBM, RJE, reservations, word processing, banking, process control, etc. |
| Presentation | Prepares data for direct manipulation by users | Display formats, code conversion, language, terminal emulation, data exchange, etc. |
| Session | Manages end user dialogue and resources | Log on, user verify, equipment allocation, token management, etc. |
| Transport | End-to-end control for session path | Msg blocking and assembly, mux, error control, sequencing, failure recovery, etc. |
| Network | Manages nodal routing | Addressing, routes over nodes, flow control, etc. |
| Data link | Node-to-node protocol | Synchronization, error control, ACK, NAK, etc. |
| Physical | Electrical/mechanical | 1553, 232C, 499, etc. |

node-to-node protocol, but there is a physical connection and addressing is present (nodal routing) through the dialing mechanism.

Figure 9-5 provides an example of the interpretation of the OSI paradigm as applied to the execution of built-in test equipment (BIT) in a system. The BIT manager issues a command to initiate BIT at the application level. The command has an address, which is noted as a network level function. In this view, there are no functions at the presentation level for issuing the command. The command is issued to the physical layer using the protocol introduced at the data link level.

The command is received at all potential destinations and decoded at the network level. The addressee then initiates BIT and determines the pass/fail status at the application level. The status message is addressed at the network level and sent through the data link and physical layers to the host. After message decoding at the network level, a status message is prepared at the presentation level and the message is then displayed at the host application level.

The ISO model has value in that it provides a structured approach to issues associated with successive levels of design. As such, it represents another

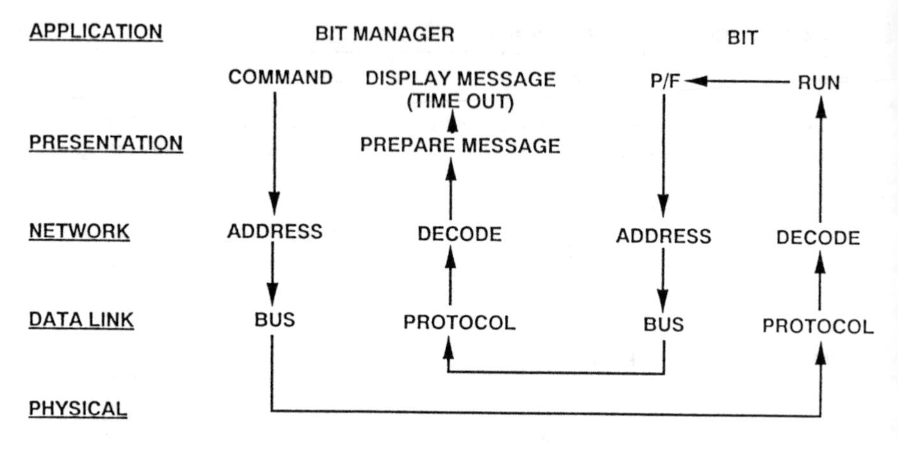


**FIGURE 9-5.** BIT example.

view that can be useful as a guide in organizing a top-down approach in conjunction with the use of $N^2$ diagrams.

## INTERFACE CONTROL DOCUMENTS

Interface requirements and designs are commonly assembled in documents called interface control documents (ICDs). There are no widely accepted standards, military or other, for the format and content of ICDs. The following guidelines, however, should be useful.

There should be one ICD for each interface. This should take the form of one document or a single chapter in a document that covers the complete interface for each of the dotted boxes that appear in your $N^2$ diagram. The document, therefore, will characteristically need to be generated and signed off by at least the two affected COGEs working in concert. The ICD should be a single repository for interface requirements and the detailed interface design. This means that it is a living document throughout development up to the critical design review freeze point. It also means that the finished document contains the detailed specification of the finished product and thus has utility as a user's manual.

An outline for the ICD is given in Figure 9-6. Section 3.1 furnishes a brief overview of the function on the interface. The level of description is similar to that provided for the top-level English descriptions at the higher levels of the $N^2$ diagram.

The actual interface requirements are divided into software requirements and hardware requirements. The software requirements are for functions

1.0 Introduction/Scope

2.0 Applicable Documents

3.0 Requirements

  3.1 Functional Description

  3.2 Interface Requirements

    3.2.1 Software Requirements

    3.2.2 Hardware Requirements

  3.3 Environmental Requirements

  3.4 Quality Assurance Requirements

4.0 Appendices

**FIGURE 9-6.** Interface control document outline.

between ISO model layers or for modules that interact within a single ISO level, above the ISO physical layer. These requirements should include data and message types, rates, transfer control mechanisms, formats, content, accuracies, response times, error detection and recovery, protocols, and all other functions that may be defined from the ISO application layer through to and including the data link layer (see Table 9-1).

The software interface description should include a clear statement of the sequence of message/data interchange that transpires between the interfacing software entities for each type of information transfer. The hardware requirements section, section 3.3.2, includes all physical electronic, mechanical, optical, hydraulic/pneumatic, and other interfaces. These include such items as the ISO communications physical layer, radios, mechanical matings, and so on. The finished ICD should contain detailed information in a clear and appropriate form, such as mechanical drawings, interfacing circuit component schematics, pin assignments, and electrical wave forms and timing. Section 3.3 covers environmental requirements. These include items such as electromagnetic compatibility, radiation, thermal, pressure, vibration, shock, and dynamic and static loading. The appendices should include additional information that may be required to clarify aspects of the interface, as well as a glossary of terms.

The ICD outline given in Figure 9-6 is not a hard standard. Rather, it is a guide that fits most cases. Modification to this structure may be adopted as circumstances require. The proposed form and format for the ICD is first produced by the systems engineer. It is then presented to the SDT for discussion, finalization, and concurrence.

**References**

1. Lano, R.J. 1977. *The $N^2$ Chart.* TRW-SS-77-04, TRW, Inc. One Space Park, Redondo Beach, Calif., November, 1977.
2. Zimmermann, H. 1980. OSI Reference Model—the ISO model of architecture for open systems interconnection. IEEE *Transactions on Communications*, Volume Com-28, Number 4, April 1980.

# 10

# Requirements and Specifications

This chapter addresses the basic approach to establishing the system functional requirements document (FRD), the system specification document (SSD), the software requirements document (SRD), and the software design document (SDD).

The FRD is the top-level system requirements document that concentrates on "what" the system must do in measurable terms. The SSD responds to the FRD, with design specifications on "how" the system shall meet requirements stated in the FRD. The SSD also allocates the implementation functions between hardware and software. The SRD responds to this allocation with a top-level description of software requirements. The SDD provides the detailed software design, using data flow diagrams and an appropriate program design language (PDL), organized with increasing levels of detail.

In large systems, similar requirements and specifications documents may be required at lower levels than the system, such as at the segment, element, or subsystem levels. Such documents follow very similar formats, but are restricted to their partitioned functions.

It is always desirable to limit the amount of documentation to no more than that which is required to effectively support the design and implementation. Lower-level documents should be used only when they are required and not simply as a matter of policy. A good guideline for this need is to determine whether there is meaningful additional material required beyond the SSD and SDD levels. Such conditions typically arise in larger in-house developments, where different organizational entities are involved or when contractors are used.

The following sections discuss the FRD, SSD, SRD, and SDD, respectively.

## SYSTEM FUNCTIONAL REQUIREMENTS

Development of the FRD represents the earliest stages in system design. Completion of the FRD implies a number of significant accomplishments. The most salient of these are the following:

- The FRD establishes the top-level commitment for the entire project effort to which all subsequent design and technical documentation must conform and be traceable to.
- User needs and system constraints have been thoroughly assessed to provide requirements that are technically realistic and at the same time will allow the user's mission to succeed.
- All system functions in terms of "what" the system must do have been stated in measurable terms.
- As systems engineer, you are prepared to freeze the contents of the FRD at the upcoming system requirements review.
- System partitioning has taken place, which consists of allocating all system functions to functional areas and/or to subsystems.

### Realism of Requirements

The process of FR construction clearly entails establishing attainable requirements. This realism may require looking forward beyond the "what" level into specific design issues, to insure that implementation is feasible. The extent of the forward look required to support the development of the final FRD is a function of the systems engineering development paradigm employed (see Chapter 2). In the staircase with feedback approach, the forward look may involve design trade-off studies. In alternative approaches, prototyping, use of the spiral model, or actual delivery of interim systems to the operational environment may be called for prior to the FRD being finalized.

The need for realism also presupposes that the finished FRD may not necessarily meet all of the user's needs as stated in the user needs document (UND). The UND is not a commitment to implementation. It is basically a complete wish list. The UND does, however, provide an important source from which the FRD is to be derived. It is the FRD that constitutes the first serious and highly structured commitment in the design and implementation process.

This point is visualized in Figure 10-1. The stated user need space is typically larger than the ultimate functional requirement space. The development of the FRD often involves paring down the full range of user "wants" to a realistic core of functions that the implementor can commit to. This does

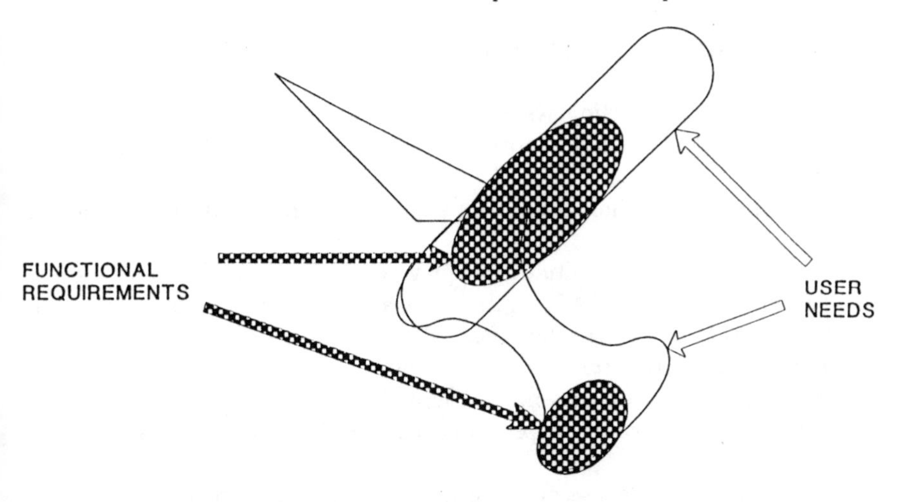


**FIGURE 10-1.** User needs vs. functional requirements.

not suggest that research and development may not be required, but it does mean that it is feasible to meet the functional requirements in a manner consistent with the system prioritized competing design characteristics, which may involve issues of cost and schedule. The SDT, with user representation, is the principle arena in which this negotiation takes place.

## Measurability of Requirements

Every requirement in the FRD will eventually need to be validated. This means, as stated above, that each requirement must be measurable. Measurability calls for the careful and complete definition of terms. Even the most common and agreed-to terms may mean different things to different people. If different interpretations are at all possible, it is imperative to discover this now rather than later during testing and acceptance exercises.

The following are examples of unmeasurable requirements:

- The personnel reporting program shall perform in real time.
- The system shall provide adequate memory for growth.
- The system shall provide a response time of 4 seconds for all non-administrative messages.
- All terminal interfaces shall be user friendly.
- The system shall provide ample HELP files.

The following are examples of measurable requirements:

- The personnel reporting system shall be capable of completing printout of individual personnel reports in 1 minute or less 90 percent of the time.
- The system design shall provide for a memory margin of 100 percent at the time of CDR.
- The response time for initiating display of non-administrative messages at user terminals shall be 4 seconds on the average and less than or equal to 20 seconds 95 percent of the time.
- Don't use the term user friendly—see text.
- The system shall provide context-oriented HELP files that enable terminal users to resolve system use procedural issues 80 percent of the time.

In the first example, use of the term "real time" is completely avoided. It is a highly relative term whose scope is completely dependent on a particular problem environment. In one application, real time can be microseconds, and in another it could be hours.

In a fighting aircraft involved in target identification, acquisition, and engagement, real time can be milliseconds, even microseconds. In a medical setting, results of routine blood tests are often received the next day. Under such nonemergency conditions, the physician assumes that appropriate actions taken within 24 hours constitute a real-time response with respect to the progress of the patient's condition. The term "real time" in itself also does not clarify the exact points in time in which real time is to be measured.

In the second example, the requirement for memory margin is restated with a measurable margin value to be realized at a specific point in time.

The unmeasurable requirement in the third example does not sufficiently define response time in terms of the beginning of a message, the end of a message, or at what location in the system the transit of the message is to be measured. The restated requirement clarifies these points and also furnishes two statistical data points that are required to characterize the service time density function in question. That is, specification of an average value alone does not fix the variance of the distribution of interest.

The term "user friendly," though popular, is not in itself measurable in any consistent way. These issues should be addressed through proper ergonomic requirements that include those for maintaining posture, lighting, display distances, angles, and so on.

Requirements for including HELP files are common, but again must be stated in ways that allow the eventual unambiguous validation that the requirement has been met.

### Conflicting Requirements

While requirements must always be realistic and measurable, they must also be nonconflicting. For example, the requirement that an armored fighting vehicle be rapidly deployable to all terrain types along with the requirement that the vehicle withstand state-of-the-art antitank ordnance may be impossible to meet simultaneously.

With current technologies, the first requirement demands a weight in the neighborhood of 20 tons and the second requires a weight in the 50-ton range. Deployability and survivability are clearly conflicting attributes. The user, however, may wish for a design that addresses conflicting issues in the best manner possible. In these cases, it is useful to state acceptable ranges for conflicting requirements, as opposed to simple absolute values. This gives the designers and implementors a clear understanding of the extent of trade-offs that will be tolerated.

One useful way to convey these limits is through the use of *attribute utility trade curves.* Four examples of such trade curves are shown in Figure 10-2. The abscissa values under the curves in the examples designate the acceptable numerical range for the requirement. The ordinate values give the relative importance, or utility, of meeting the requirement with a value in the numerical range.

The interpretation of the MTTR example in the figure is that 5 points will be awarded to the utility for an MTTR of 2 hours. The utility linearly decreases to a maximum value to be tolerated for MTTR of 5 hours where the utility is zero.

The CEP example establishes that the tolerated range for the *circular error probability* is from 10 to 30 feet. Attainment of a CEP of 10 feet is awarded 10 utility points. Attainment of a CEP of 20 feet is also highly acceptable. It is not until the design results in a CEP of over 25 feet that the utility rapidly descends. A different utility curve characteristic is shown for weapon lethality.

Attribute utility trade curves can also be used to convey the relative importance of programmatic issues as well. The schedule example in Figure 10-2 conveys that the system has no utility if it is delivered later than 18 months from the start. The curve also indicates that there is no increase in utility if the system is delivered at any point earlier than 18 months from start.

When the utility scheme is used, the FRD includes an overall system utility value that is to be met. The individual utility curves provide the implementors with a precise understanding of the extent to which design trade-offs can be made to meet the system utility metric. They also provide a means of quantifying the extent to which vendors propose to meet the requirement mix when evaluating bids and proposals.

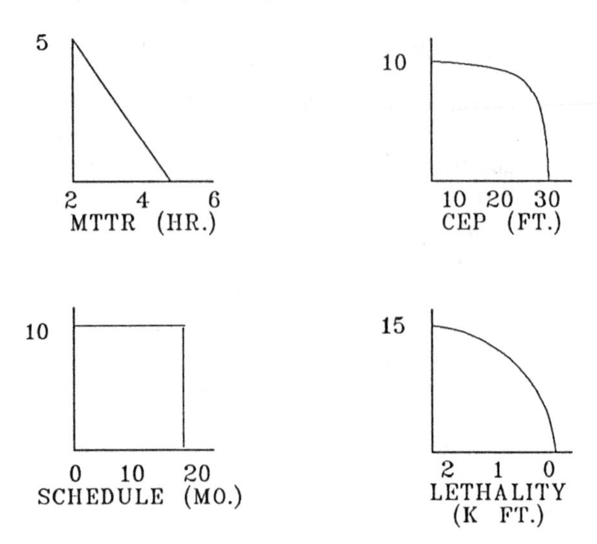


**FIGURE 10-2.** Attribute utility trade curves.

## System Partitioning

Construction of the FRD includes the defining of system functional areas. This requires system partitioning, which consists of allocating system-level functions to functional areas. Functional areas refer to system elements and/or subsystems. This process involves the structured decomposition of system functions into a hierarchy of lesser functions whose sums combine to meet all system requirements.

There is no widely accepted standard for developing such hierarchies. Approaches to decomposition vary among organizations and among individuals within organizations.

One helpful guideline is the generic Department of Defense (DOD) system hierarchy, as shown in Figure 8-6. A first example of the use of this hierarchy is given in Figure 10-3 for an infrared space instrument.

The system is first divided into a set of segments. Segments are the top-level functional entities of the system. In example of Figure 10-3, they consist of the entire flight segment and the entire ground segment. Note that the rationale for this particular segmentation is basically location. The flight segment is then decomposed into three elements: the spacecraft, the launch vehicle, and a support element. The spacecraft itself is composed of subsystems, examples of which are the instrument payload, communications, the structure, and attitude control. The payload is further subdivided into components of the infrared instrument, consisting of the instrument sensor,

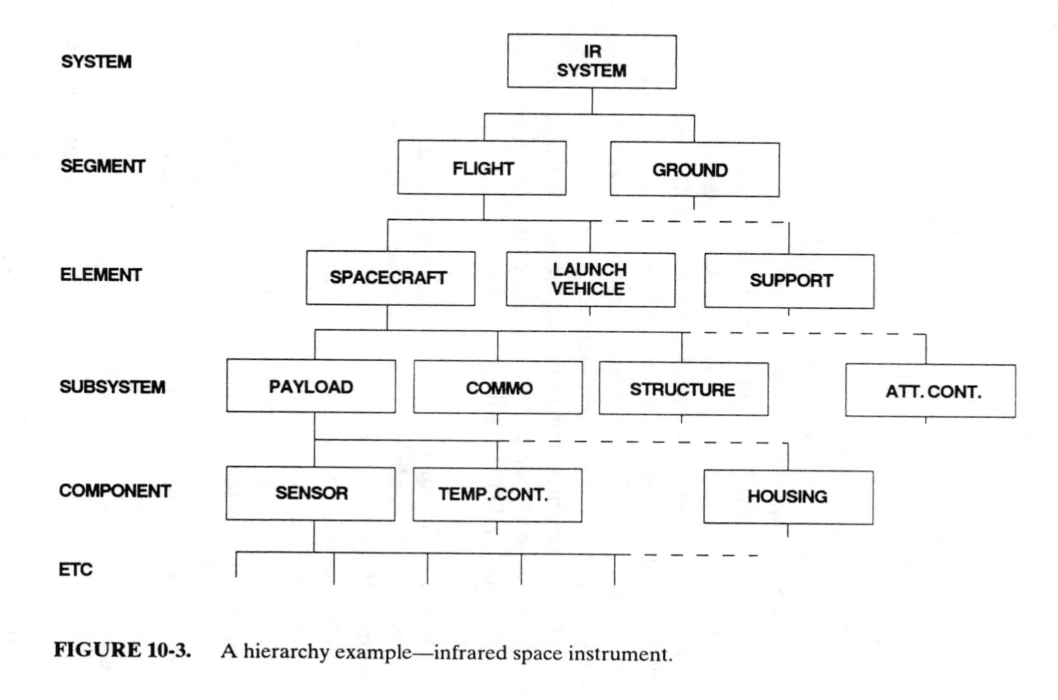


**FIGURE 10-3.** A hierarchy example—infrared space instrument.

temperature control, housing, and so on. A similar top-down decomposition of the ground segment takes place.

Figure 10-4 provides a second example of functional decomposition for a locomotive. In this case, the segments consist of power, traction, control, and the structure. The power segment is broken down into elements consisting of the engine, engine intake, engine exhaust, lubrication, and the starting, cooling, and fuel elements. The traction segment provides the interface between the engine and the track surface through such elements as gears and truck assemblies. The control segment provides for all mechanical and electrical system controls. The structural segment includes such elements as the underframe, the cab, and hood structures. Unlike the first example, this segmentation is based upon function.

It is not absolutely imperative that one adhere to the terms segment, element, subsystem, and so on, in performing partitioning. In the locomotive example, one may wish to drop the term segment so that the power unit becomes an element and the engine, intake, exhaust, and so on become subsystems. This is perfectly reasonable. In the IR instrument example, however, it is convenient to use segments at the flight and ground system levels. In any case, the DOD hierarchy provides a useful starting point for structuring the partitioning of any system.

In the above examples, the initial segmentation was based on physical location in one case and on function in the other. Table 10-1 lists these and other possible rationale for the decomposition of systems from a given level to the next.

Selecting any of the partitioning rationales at the segment level has direct consequences at all lower levels. One consequence of the air/ground segment breakdown of the IR system is that a communications subsystem will be needed at some point in both segments. Clearly, the air and ground segments must communicate with each other. In actual fact, the complete communications function between the segments would likely be under the direction of a single cognizant engineer and treated in the work breakdown structure as a single deliverable subsystem. Thus, while the rationale for segmentation is physical location, the motivation for subsystem definition in the same system is functional. There is nothing inherently improper with this, although such choices can often confuse the casual observer.

The initial system decomposition is an important exercise, with lasting effects on the organization of all future work. It is akin to laying down a bowling ball. There will be no difficulty in explaining different partitioning rationale at different levels if the process has been well thought out and logically conceived.

The decomposition exercise is directly related to the process of arriving at the work breakdown structure for the mission product discussed in

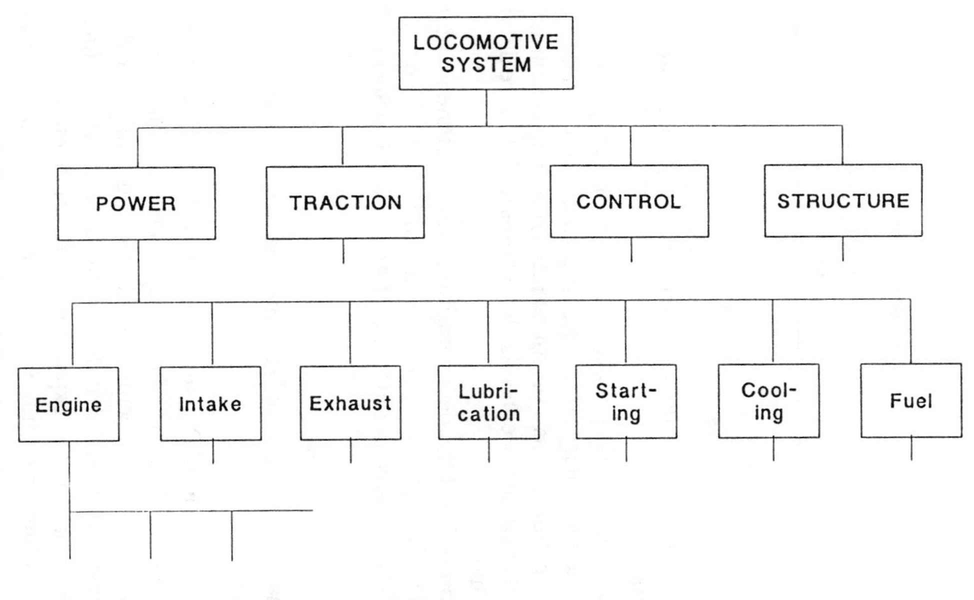


**FIGURE 10-4.** Locomotive system partitioning.

**TABLE 10-1 Rationales for System Partitioning**

| Partition By | Examples |
|---|---|
| Engineering discipline | Hardware, software, VSLI, graphics, etc. |
| Physical location, environmental separation, etc. | Air, ground, space, moon, geography, tactical |
| Function | Communications, fire control, navigation, attitude control, etc. |
| Politics | Nontechnically driven, international consortiums, etc. |

Chapter 5. The processes influence each other and may often result in identical structures.

During the decomposition process, the systems engineer should also constantly pursue the concept of developing simple and well-defined interfaces. System partitioning has a profound influence on the eventual detailed definition of interfaces at lower levels. Procedures for interface definition are presented in Chapter 9.

The major outcome of the decomposition exercise at the point in time at which the FRD is being formulated is the identification of functional levels of the system. These levels should be treated specifically in Section 3.7 of the FRD.

## The Functional Requirements Document (FRD) Form and Format

A generic outline for an FRD is shown in Figure 10-5. The outline is similar in structure to that of MIL-STD 490, type A Specification Practices, which, while being designed for specification documents, still provides a good checklist of items to be covered at the functional level. The outline is meant to be a guide. Sections may be expanded, added, or deleted as needed.

The System Definition section provides an overview of the system function and its mission with supporting diagrams. Functional areas, or subsystems, are defined as a part of this definition. Requirements for the functional areas themselves are covered in Section 3.7. System-level interfaces and interface requirements are described in terms of specific inputs and output products.

Operational and organizational concepts refer to the relationship of the

1.0 Scope

2.0 Applicable Documents

3.0 Requirements
- 3.1 System Definition
  - 3.1.1 System Function
  - 3.1.2 Mission Descriptions
  - 3.1.3 System Diagrams
  - 3.1.4 System Interfaces
  - 3.1.5 Operational and Organizational Concepts
- 3.2 System Characteristics
  - 3.2.1 Performance Characteristics
  - 3.2.2 Physical Characteristics
  - 3.2.3 Reliability
  - 3.2.4 Maintainability
  - 3.2.5 Availability
  - 3.2.6 Environmental
- 3.3 Design and Construction
  - 3.3.1 Security
  - 3.3.2 Safety
  - 3.3.3 Human Engineering
- 3.4 Documentation
- 3.5 Logistics
  - 3.5.1 Facilities
  - 3.5.2 Transportation and Handling
  - 3.5.3 Personnel and Training
- 3.6 Prioritized Competing Design Characteristics
- 3.7 Functional Area Characteristics

4.0 Quality Assurance Provisions

5.0 Appendices

**FIGURE 10-5.** Functional requirements document outline.

system to the overall mission, how its operation accomplishes or supports the mission, characteristics of the direct users of the system, the system support mechanism concepts, and the structure of the overall user organization. Section 3.1 should not be verbose or lengthy. It represents a top-level, self-contained, and comprehensive system description. It should be crisp, terse, to the point, and complete.

Performance characteristics include such items as acceleration requirements, fuel economy, message types, message formats, response times, data volumes, error rates, system life, storage requirements, turning radius, propagation coverage, ordnance delivery, and so forth. These are measurable items directly related to performance to be observed at the system boundary.

Physical characteristics cover weight limits, dimensional limitations, crew space requirements, ingress, egress, maintenance access, and so on.

System-level RAM requirements are covered in Sections 3.2.3, 3.2.4, and 3.2.5. At the functional requirements level, it is common to state the system availability requirement only. This allows design trade-offs between reliability and the maintenance support mechanism to be made at the specification (design) level of effort, thus providing maximum freedom to the designers. In rare cases, it may be necessary to state reliability or maintainability requirements at the functional level. Bear in mind, however, that once two of the three system RAM parameters has been set, then all three have been set.

Environmental requirements include those for wind, rain, temperature, shock, motion, noise, electromagnetic, over-pressure, blast, humidity, radiation, chemical, biological, nuclear, and so on. These describe the external conditions that the other system characteristics will be subjected to.

The design and construction sections under Section 3.3 include additional requirements for such items as security, safety, and human engineering.

A system is not a system without documentation. Section 3.4 states all requirements for documentation to be delivered with the system.

The logistics section covers any known functional requirements not already covered related to facilities, system support, transportation and handling, personnel, and required training levels.

Section 3.6 contains a listing of the system prioritized competing design characteristics, along with a description of their meaning and rationale for prioritization. This section can generally be taken directly from the system engineering management plan (see Chapter 6).

Functional area characteristics are given in Section 3.7. There should be one subsection for each subsystem covering performance and physical characteristics as a minimum.

Section 4.0 addresses implementation constraints. This is the only portion of the FRD that states "how" any part of the implementation is to be achieved. The need for this at the FRD level may arise if, for example, a communication system must interface with existing computers such that a particular protocol must be used at given system interfaces. Systems that are undergoing upgrades will commonly have constraints imposed upon them. Even brand new systems often need to make use of, or interface with, existing equipments. This section allows the authors of the FRD to instruct the implementors on these types of limitations.

## THE SYSTEM SPECIFICATION DOCUMENT

An outline for the system specification document (SSD) is given in Figure 10-6. While the structure is very similar to the FRD, the content provides significantly more detail. As with the FRD, the outline is presented as a guide to assist the SSD developer in considering all pertinent factors.

The important major difference between the two documents is that the FRD basically states "what" the system must do and the SSD states specifically "how" the system must meet the functional requirements.

For example, where the FRD might state a requirement as:

"The system shall provide for storage of all institutional personnel records over the period of the system life,"

---

1.0 Scope
2.0 Applicable Documents
3.0 Requirements
    3.1 System Definition
        3.1.1 System Description
        3.1.2 System Diagrams
        3.1.3 System Interfaces
    3.2 System Characteristics
        3.2.1 Performance Characteristics
            3.2.1.1 Hardware
            3.2.1.2 Software
        3.2.2 Physical Characteristics
        3.2.3 Reliability
        3.2.4 Maintainability
        3.2.5 Availability
        3.2.6 Environmental
    3.3 Design and Construction
        3.3.1 Security
        3.3.2 Safety
        3.3.3 Human Engineering
    3.4 Logistics
        3.4.1 Facilities
        3.4.2 Transportation and Handling
        3.4.3 Personnel and Training
    3.5 Functional Area Characteristics
4.0 Quality Assurance Provisions
5.0 Appendices

---

**FIGURE 10-6.** System specification document outline.

the SSD would combine this requirement along with requirements for the size of personnel records, the number of records (including requirements for system growth during the system lifetime), and the response time for the recall of records to develop a specific requirement for how the storage shall be provided, such as:

"The system shall provide three megabytes of on-board RAM and 10 megabytes of disk storage to accommodate institutional personnel records over the system lifetime of five years."

Similarly, interface flow discussed in Section 3.1 of the FRD is stated in English terms using $N^2$ diagrams, while the SSD discussion is in terms of specific formats to be used. In the same manner, the RAM section in the FRD covers the top-level system requirements. The SSD in turn responds to these requirements with a complete specification for allocating reliability and logistics support mechanisms to achieve these requirements.

A request for proposal (RFP) should generally leave as much design freedom as possible for the prospective vendors. The FRD level of detail is thus preferred for an RFP with an appropriate constraint section. A request for quotation (RFQ) is a request for quotes on a specific detailed configuration or piece of equipment. RFQs are generally written using the SSD level of detail. As a rule, when an RFQ is issued, an FRD has already been generated and baselined.

Depending on the size and complexity of the system under development, subsystem specifications may be included in Section 3.7 of the SSD or may warrant their own documents. A separate subsystem specification document would typically be written if an RFQ for a subsystem were to be issued. The form and format follows the SSD specification with the system now replaced by the subsystem. However, enough information is included at the system description level to give the prospective vendors a clear understanding of the subsystem's role in the overall system. In this case, it may be acceptable to repeat all or parts of Section 3.1 of the FRD in the subsystem specification. In general, however, simple repetition of document content in lower-level documents should be avoided to minimize the complexity of incorporating changes at a later time. The chore of updating documentation in a complete and consistent fashion can be greatly alleviated through the use of computer-based documentation aids.

All of the requirements stated in design documentation subsequent to the FRD must be traceable to their parent document in both forward and backward directions. This is the process of verification, which is an integral part of configuration management, as discussed in Chapter 8.

## THE SOFTWARE REQUIREMENTS DOCUMENT

The content of the SRD is similar in level of detail to the FRD in that it addresses "what" the software must do in measurable terms. The specific requirements covered in the SRD are defined by the SSD where the meeting of functional requirements is allocated to hardware and software.

Numerous standards exist for the formats for software requirements and specification documents. The outline given in Figure 10-7 for the SRD is a simplified composite of a number of these (reference IEEE, 490 B5, and Pressman) and serves as a practical guideline.

Section 2 provides a top-level narrative description of the overall software problem, with supporting top-level data diagrams and structured English. The description includes information content and structures, including data base structures. Also implicit in the data flow description is a complete narrative description of software system interfaces. Design constraints, such as restrictions imposed by the target environment, may also be included in this section.

1.0 Introduction
- 1.1 Purpose
- 1.2 Applicable Documents
- 1.3 Definitions/Acronyms

2.0 General Description
- 2.1 Problem Definition
- 2.2 Data Flow
- 2.3 Informatoin Content
- 2.4 Information Structures
- 2.5 Design Constraints

3.0 Functional Partitioning
- 3.1 Program A
  - 3.1.1 Process Narrative
  - 3.1.2 Performance Requirements
- 3.2 Program B

.

.

.

- 3.X Program X

4.0 Test Requirements

5.0 Appendices

**FIGURE 10-7.** Software requirements document outline.

The SRD also partitions the software in a fashion similar to the original system partitioning that takes place in the FRD. A proven method for software partitioning is the use of structured analysis techniques with data flow diagrams (see Chapter 15).

Section 3 partitions the overall software problem into programs that become computer program configuration items (CPCIs). Narrative descriptions for each functional program and performance requirements are supported by lower-level data flow diagrams. All data flow diagrams must

1.0 Introduction
- 1.1 Purpose
- 1.2 Applicable Documents
- 1.3 Definitions/Acronyms

2.0 Design Description
- 2.1 Program Structures
- 2.2 Data Description
- 2.3 File Descriptions
- 2.4 Data Bases
- 2.5 Design Constraints

3.0 Functional Partitioning
- 3.1 Program A
  - 3.1.1 CPC a
    - 3.1.1.1 Process Narrative
    - 3.1.1.2 Design Language Description
    - 3.1.1.3 Data Organization
    - 3.1.1.4 Test Requirements
  - 3.1.2 CPC b

    .
    .
    .
- 3.2 Program B

  .
  .
  .
- 3.X Program X

4.0 Test Requirements

5.0 PDL Representation

6.0 Source Code List

7.0 Appendices

**FIGURE 10-8.** Software design document outline.

include consistent interface descriptions and terminology where interfaces between CPCIs exist. All performance requirements must be measurable.

Section 4.0 covers software test requirements, which must account for testing of all performance requirements. Test bounds, types, and classes are defined along with anticipated results.

## THE SOFTWARE DESIGN DOCUMENT

An outline for the SDD is given in Figure 10-8. The SDD is constructed in parallel, with the design process successively making use of products of structured analysis followed by structured design.

Section 2.0 makes use of data flow diagrams that are products of the preliminary software design activity. Program structures and program interfaces are reviewed and refined. Included are complete descriptions of data types, files, and data base structures and assignments of global data.

Section 3.0 emerges as the detailed design proceeds. Programs are decomposed into computer program components (CPCs), modules, and procedures. CPC definitions begin with a narrative description followed by successively detailed data flow diagrams at the CPC and module levels and are finally represented using a program design language (PDL). Test requirements for each CPC are included with details at the module level as required.

Section 4.0 refines test procedures at the program level.

Section 5.0 assembles a complete PDL representation of the software at a single location. This representation should ideally begin with a single-page PDL description of all of the software under development. The next level may use three to five pages, and so on. PDL descriptions are organized in successive levels of detail around the software decomposition of CPCIs, CPCs, and modules.

When the system acceptance testing is completed, a source code listing is finally included as Section 6.0. This step insures that a complete design description of the software is provided as a part of the SDD.

When constructed properly, the SRD and SDD should provide a complete and highly structured documentation set, such that readers unfamiliar with the product can clearly find their way from a top-level understanding of product goals to the lowest levels of detailed coded implementation.

# 11

# Integrated Logistics Support (ILS)

Logistics support addresses all issues associated with the execution of support activities required for both the development and operation of the mission product. In addition to the maintenance plan, logistics support consists of designs and procedures for the items listed in Table 11-1.

It is evident that the systems engineer must consider two distinct logistic support strategies—one for the sole support of development activities and another to support field operations. The former is generated to support the design and implementation process. The latter is a part of the integrated design process for support of the mission product after delivery to the user(s).

The first step in developing any ILS plan is to carry out a detailed *logistics support analysis* (LSA). MIL-STD-1388-1 provides an excellent checklist for the items that must be addressed in carrying out a comprehensive LSA. The LSA results in the definition of concepts and detailed approaches to each of the major ILS items.

These operational support strategies can have a profound impact on system design and on system costs trade-offs. Systems that are difficult to

**TABLE 11-1 Major Elements of Logistics Support**

| |
|---|
| Supply support |
| Test equipment |
| Transportation and handling |
| Technical data packages |
| Facilities |
| Personnel and training |

troubleshoot and restore upon failure may require highly trained personnel and test facilities in order to maintain availability requirements. These designs increase operational costs. Alternatively, support concepts based on self-diagnosing modules that are easily replaced can significantly alleviate requirements for training, special test equipment, and complex spare inventories in the operational setting.

Logistics support concepts are first defined prior to the systems requirements review. The development of these initial concepts involves major decisions that have direct impacts on the overall system design. As with all design decisions, assistance in the making of major ILS decisions can be effectively guided by consistent review of the prioritized competing design characteristics for the system under analysis.

The LSA progresses to a preliminary ILS plan prior to the preliminary design review and finally matures to a complete ILS plan that is presented for detailed examination at the critical design review. The following paragraphs discuss the major elements of logistics support.

## SYSTEM MAINTENANCE PLAN

The *system maintenance plan* (SMP) follows directly from the LSA effort. Reliability and maintainability analyses are conducted to support the SMP, both for development and for operations. *Life cycle cost* considerations are an important part of the design of the operations SMP. Over the system lifetime, operational costs are typically greater than development costs.

The issue, however, can be complicated. It is a peculiarity that in some large systems the fundings for development and for operations come from different costing centers within the same procuring agency. In such instances, motivation to minimize development costs can actually take precedence over life cycle cost considerations, such that higher total lifetime costs result.

There are other instances where the implementor does not directly assume maintenance costs. In the development of commercial products (cars, appliances, locomotives, etc.), where the consumer typically assumes maintenance responsibility, it is wise to consider maintainability as a selling point. Still, in the marketing of less expensive products with relatively short lifetimes (calculators, fashion watches, etc.), a valid approach may be that maintenance is replaced by simple unit exchange. In some cases, manufacturers have found it cost-effective to go one step further by dropping their quality control function all together, marketing a higher ratio of defective units, and letting the consumer, in effect, carry out quality control by returning defective units for exchange. This approach, however, is not in keeping with modern quality management trends that place a premium on customer satisfaction.

In most systems of any reasonable size, however, it is almost always expedient to implement designs that reduce operational costs when possible. One common approach to the reduction of operational costs is in the use of built-in test equipment (BITE). BITE consists of special test equipment that is built into a functional unit and is usually executed by special test software. The addition of BITE capability, while increasing development costs, is typically employed to reduce maintenance costs. The use of a built-in test (BIT) software is an effective way to reduce the time for fault isolation and hence the requirements for training in forward service areas. Because it can reduce the *mean time to restore* (MTTR), it can also provide relief on requirements for module reliability (see Appendix B).

The design of BITE and BIT involves the modular organization of hardware into units that can quickly be diagnosed and replaced. These units are often referred to as line replaceable units because they are replaced at the "front" line of the operational setting. In this support scheme, faulty LRUs are removed and shipped to an intermediate support area where repair and re-calibration can take place. In the military, these intermediate areas are often highly mobile and limited. Units that cannot be repaired at this level are returned to the Depot level, where complete repair capabilities are available.

The analysis that supports the design of such three-tiered systems is not so concerned with the fate of a single LRU as it is with maintaining inventory flow up and down the system. Required inventories at each of the forward, intermediate, and depot levels are determined by LRU and system availability calculations and/or experience.

Operational maintenance planning involves defining maintenance concepts to support the mission product in operations. It entails considering each of the elements listed in Table 11-1 as well as such factors as safety, specialty engineering, system availability requirements, manufacturing, customer service, and, of course, cost trade-offs. At a minimum, the maintenance plan should contain one section for each element in Table 11-1 and a separate section describing how these elements are integrated to provide maintenance procedures. These procedures should result in effective system support at all operational levels that support the meeting of system availability requirements.

System development maintenance planing similarly entails consideration of the elements listed in Table 11-1, such that the design and implementation process is efficiently supported.

Maintenance planning, of course, draws most heavily on reliability, availability, and maintainability (RAM) analysis. The fundamentals of RAM analysis are discussed in Appendix B and should be thoroughly reviewed prior to developing the maintenance plan.

## SUPPORT EQUIPMENT

In the development domain, support equipment refers to those level 4 WBS equipment items, including tools, spare units, spare parts, inventory levels, and consumable items required for system development and maintenance of the system or portions of the system under development.

Operational support equipment categories are similar, but are identified and/or designed as part of the development effort to provide eventual operational support for the fielded system. The locations, types, and quantities of support equipment will be different for the development and operational environments. In either case, they are defined by the LSA effort and the SMP.

Equation 24 in Appendix B provides a useful means for estimating the number of spare units required at a given site to support the maintenance function.

## TEST EQUIPMENT

Test equipment in the development setting includes electronic test equipment, instrumentation, diagnostics and diagnostic equipment, mock-ups and special test rigs for fatigue, environmental and performance testing for unit, integration, systems, and acceptance testing during development. This also includes all contractor support and consumables associated with all phases of development testing.

In the operational setting, test equipment includes electronic, diagnostic, and other test equipment to be used in the field by maintenance personnel. If a multi-tiered operational support philosophy is used, test equipment applicable to each level is required. Test equipment requirements are defined in the SMP.

## TRANSPORTATION AND HANDLING

Perhaps one of the more informative incidents related to the failure to adequately consider impacts of transportation and handling on system design deals with a system developed for the shuttle cargo bay, implemented in California by experienced engineers, that wound up going to Cape Kennedy by rail and then by barge down the Mississippi because it was discovered too late that it wouldn't fit through a C5-A door. Another incident of interest involved a group that made arrangements for delivery of a number of large antenna towers in Texas for propagation testing of a prototype system during development. The engineering details were extremely well planned, but, when the crew got to the gate with three big trucks, the security guard had never heard of them. By the time security determined who the strange visitors were, the remainder of the day and half of the following day had passed, which the crew spent happily around the pool at the local Holiday Inn.

Transportation and handling (T&H) has to do with every aspect of equipment storage, packaging, preservation, and the total logistics of moving it from one place to another. The impact of T&H on design can be anywhere from minimal to significant, but there is almost always some impact, whether the product is consumer-oriented or a high-tech military system.

It is quite common for acceptance testing to take place after installation. This means, by definition, that your entire system will have to be picked up and moved before the customer signs off. This move is still technically in the development phase. It is not unusual for projects to suddenly consider these aspects of design well into the design phase—dangerously late. On-site prototyping and feasibility testing of subsystems as well as customer training exercises may also be a part of the development plan that may require further T&H considerations.

Where operational mobility is a system requirement, T&H will clearly impact a design. Here the concern is with the ability to rapidly pick up, package, and move the product through all vibrational and environmental conditions to be encountered and to reestablish operations at a different location. Early and thorough determination of design impacts on size, weight, mountings, and so on can avoid expensive work-arounds later. The placement of hooks at centers of gravity or other strategic locations, accesses for fork lifts, handles, wheels, and slides can all be unfortunate appendages when designed as afterthoughts. Even when operational mobility is not a requirement, equipment still has to be designed to ship, be returned, and be shipped again without harm.

## TECHNICAL DATA PACKAGES

This is the site in the generic WBS where the responsibility for coordinating all technical documentation resides. In development, this is the point of coordination for all user needs, requirements, specifications, interface, management plans, options analysis, special study, development ILS, configuration management, testing, hardware and software design documentation, and anything else that supports development. Development technical data packages are typically not delivered to the customer.

Operational technical data packages are deliverables. They typically consist of instructions for assembly, installation, trouble shooting, operations manuals, user manuals, maintenance manuals, training manuals and equipment, and technical functional descriptions of hardware, software, system, and ILS operations.

When constructing operational support documentation, it is important to make a clear distinction as to the aim. One aim is to provide a reference document. Reference documents are excellent for people who already understand the basic product mission and a fair amount of detail regarding hardware, software, and operations. Reference documents are of absolutely no value as learning aids. They serve principally to remind the user of forgotten items or to clarify procedures the user is already familiar with. (Dictionaries don't teach English, nor are they intended to.)

Another distinct aim of documentation is to teach. Teaching documents are totally different from reference documents in that they take the reader through a clear top-down descriptive process that only gradually comes to significant detail. Teaching documents are typically very difficult for engineers to write. When an author knows a lot about something, it is very easy to take fundamentals for granted and to wind up instructing at a level inappropriate to the student audience. Software documentation is notorious for this. Most engineers have not had formal teaching experience, nor are they motivated in these directions. Good technical people, by nature, like to get on with what they are doing—to keep pace with technology and with their colleagues in pressing for solutions. Stopping work to generate documents to help people who may be months or even years behind them to understand basics is an incredible imposition on them.

Technical writers, on the other hand, do this kind of thing by trade, and a good one is well worth the time and money in meeting schedules for producing training and learning material. A common argument against using technical writers is that they "don't know the system" or "the learning curve will take time." The fact is that unfamiliarity is a distinct advantage, in that it forces treatment of the subject from square one—exactly what is desired in a training (learning) document. Good technical writers have broad technical backgrounds and writing skills in which engineers are notoriously weak. Technical writers can be particularly useful in generating documents designed to instruct.

A third distinct type of documentation is design documentation. In some cases, design documentation or portions of it are used to support technical manuals delivered to operations. Problems in generating this kind of documentation for mechanical and electronic hardware items is generally not too severe because designers need layouts, drawings, string lists, logic equations, topological illustrations, or schematic representations in one form or another to do the job. It is a natural part of the lore in constructing the physical.

This is not true in programming. A programmer can go to his or her terminal and simply start coding without any requirements or design documentation—and you can count on them doing exactly that. They "design" by "coding" because this is the most fun.

A few years ago, I had one programmer threaten to quit rather than generate his own design documentation—forget reference or training manuals. Now I realize that the problem was of my own doing. I was asking someone to do something that he had no professional interest in doing. Nor did he have the capability. No one likes to do something they don't know how to do. Further, I had no direct control over his paycheck in the matrix organization, and the line management would not consider replacing him. His prodigious ability to produce working code for perplexing problems was unparalleled. He was simply perceived as being too valuable. "Do it or else" was not a viable solution.

All systems engineers must address the problem of software design documentation. In the end, the only way that programmers can be motivated to document requirements and design as they go is to employ a methodology that is accepted by all as an advantageous way to proceed. If there is no clear advantage in the implementation approach to integrating software documentation into the design process, programmers will always tend to document last, and then only with considerable effort and cajolement. The advantage must be to the programmer, not simply to the systems engineer who needs to somehow wrestle design documentation from those whose true interests lie elsewhere. Probably the best current methods to achieve this goal employ the proven concepts of structured analysis and design (see Chapter 13).

If the system you are implementing is of any size or complexity, you will probably be wise at this stage of your planning to use:

1. Technical writers for documentation that will instruct.
2. Technical writers and possibly cooperative implementors for reference documentation.
3. Implementors for producing design documentation, providing that you can devise or select a methodology that is advantageous for them to use. Structured analysis and design techniques provide significant advantages in developing good documentation in parallel with software development.

## FACILITIES

In development, facilities include constructing, acquiring, or converting buildings, spaces for laboratories, prototyping, training, and testing, as well as utilities, environmental control, and so on. This is also the point in your WBS to include any furniture, copiers, fax machines, telephones, lamps, secretarial personal computers, vehicles, perishables, and other items needed to support development that a given project may need to procure in addition to available existing equipment.

Operational facilities include developing, acquiring, or converting structures, vehicles, and so on that have not been specifically designated in the "auxiliary equipment" category of the mission product. A guideline for differentiation between these two categories is that auxiliary equipment is generally oriented toward supporting a mission product item or subsystem, while operational facilities refer to system-wide support facilities. A mobile field system may include a number of subsystems, each of which is mounted on its own vehicle. Thus, a vehicle on which a communications center subsystem is mounted may be classified as auxiliary equipment to that subsystem, while another vehicle that provides service to all subsystem vehicles may be classified as an ILS facility. There are always options, but consistency is to be valued. The important thing is to think of everything and to recognize that, in a good generic WBS, everything has a well-defined place.

I can recall an incident while working as a communications consultant for a public safety agency and assisting, among other things, in facility design.

"Don't forget the chairs," I was warned.

"The chairs?"

"Yeah, the last time we did this we were so hung up in communications and consoles that we forgot the chairs. Did you know that the chairs we use in the command and control center cost close to $500 a piece and we use thirty of them? That's a lot of money for us to forget. Don't forget the chairs."

Errors of omission when considering facilities for support of development and operations are among the most common.

## PERSONNEL AND TRAINING

On the development side, this item basically covers the time and money required to acquire and train new personnel. It can easily take 2 to 3 months and may require up to 8 months before newly acquired personnel are totally up to speed in their productivity. If new personnel are required, this ILS item will affect both costs and schedules.

The design of training for operations personnel can have at least two aspects. In some systems, initial training in operations and maintenance at the time of system delivery is all that is contracted for. In others, such as in the maintenance of consumer products, design of a sustaining training capability is required as part of the system ILS design. Although related, the requirements for these two approaches differ in depth. The goals of the training component of ILS need to be well articulated, as does everything else, early on when establishing system requirements. Training for operations includes training services, equipment, hardware and software aids, test equipment, and parts sufficient to teach the necessary skills for operating and maintaining the mission product. Included is the development of derived requirements and design and implementation for all training equipment, as well as the execution of the training function.

A final word on ILS. There is a simple realization that the systems engineer should adopt with regard to ILS. Before final acceptance of any system by the user, all facets of ILS will take place—*even if planning for ILS is totally disregarded*. They happen because they have to happen. The only question is: How well did the systems engineer anticipate and plan for the myriad of issues involved? Every ILS factor will come up sooner or later because all systems need support of one type or another. If you have not planned for them, even down to the logistics of arriving at a gate with a delivery, then you will need to assume a reactive mode involving adjustments and afterthoughts.

One of the purposes of systems engineering is to avoid surprises. Each of the ILS factors must be addressed carefully for their applicability. The life of a systems engineer includes few guarantees. But there is at least one: Each minute detail of logistics support that you do not find in the beginning will find you in the end.

# 12

# Trade-off Analysis

Trade-off studies are conducted throughout the systems engineering process, from initial consideration of architectural and technological options to detailed parts selection studies.

The basic components of a trade-off analysis consist of:

1. Definition of objectives;
2. Determination of selection criteria; and
3. Analysis and evaluation of alternatives.

It is a primary responsibility of the systems engineer to take the lead in identifying meaningful trade-off alternatives and to guide the methodic discharge of required studies to a productive and timely conclusion. This definition and execution of trade-off studies is a principal activity of the SDT under the leadership of the systems engineer. The nature of trade studies may vary, from architectural and system performance analyses early in the process, to evaluating technology alternatives during detailed design, to weighing of proposed design changes late in the overall process.

The definition of the objective of a trade-off study should be crisply stated in one or two sentences. The objective must be measurable so as to provide a sound basis for decision and a well-defined ending point. The measurable aspects are based on the ability of the proposed options to meet functional requirements and/or specifications.

In conducting such studies, it is highly desirable to employ study and evaluation techniques that are as objective as possible and that have a consistent basis for decision making throughout the development process. A consistent criteria for evaluating alternative courses of action throughout the systems engineering process is the list of prioritized competing design characteristics (PCDCs). The definition of PCDCs establishes a common view of top-level priorities for both the users and the developers. The use of PCDCs greatly enhances the ability of the systems engineer and the SDT to maintain a consistent and straightforward strategy for decision making, from the initial development of requirements to system transfer to the customer.

This chapter first covers the development and use of PCDCs. The range of modeling tools used throughout the systems engineering process to evaluate design concepts and trade-offs is then presented with a system architecture definition example. This example highlights how modeling techniques can be used to productively reduce the number of potential design alternatives. Basic techniques of risk analysis commonly employed to support the evaluation of alternatives are also introduced.

## PRIORITIZED COMPETING DESIGN CHARACTERISTICS

Systems are frequently conceived, designed, implemented, and delivered without a clear and consistent set of priorities in the minds of all parties involved. It is important to devise and negotiate a set of priorities directly with the customer or through effective market research so that the major factors that will influence the establishment of requirements and guide potentially crucial design decisions are clearly understood by all parties in advance. Setting priorities is also of value in achieving a common understanding of how and why initial designs or even requirements may need to be altered, should it be discovered that compromises need to be made.

Establishing clear priorities further provides a framework for the systems engineer's own consistent behavior before the SDT, management, users, and all others associated with the project. Establishing consistency throughout the entire process is a principal goal of systems engineering. Incorporating agreed-upon priorities sets the tone for that consistency at the highest level for all involved parties.

Priorities are derived primarily from listening to the "customer." They are also balanced by an engineering knowledge of what is achievable at an acceptable level of risk. The customer includes the ultimate system users, system operators, maintenance personnel, supervisors, and all levels of management—everyone who touches or is touched by the system. Typically, they all have different priorities.

An important question to ask the customer community at all levels when defining user needs is "What do you want most out of this system?" or "What aspect of this system is most important to you?" Driving issues such as cost, schedule, or performance generally surface in response to such open-ended questions if they are truly foremost in the respondent's mind.

Should clear top-level patterns not emerge, further specific probing on issues such as cost, performance, schedule, availability, flexibility, maintainability, operability, transportability, and mobility should be initiated.

## Example 1—PCDCs for the AMDARS System

Table 12-1 lists a set of PCDCs that might apply to the AMDARS system discussed in Chapter 5. The example is designed to provide a feeling for the level of consideration that goes into formulating PCDCs. In this example, it is assumed the following rationale for the ordering of PCDCs resulted from interviews and discussions during the previous development of user needs.

The system must be easily maintained. The customer desires that false readings from remote units that exceed a range of 10% be detectible. It is also desirable that a high degree of training not be required of personnel charged with diagnosing and repairing malfunctions. Finally, it is of great importance to the customer that field units be replaceable by a single individual, without leaving the cockpit of the service vessel.

The customer will subordinate system end-to-end performance in terms of sensor accuracy for the maintainability features described above. Putting it simply, a knowledge of system status and ease in handling the system is more important than the system accuracy.

The customer has specific locations in mind for installation of remote units. The ability to cover that area in the final system, however, is subordinate to

**TABLE 12-1 Example of Prioritized Competing Design Characteristics for the AMDARS System**

| | |
|---|---|
| Maintainability | Ease of equipment failure detectability, diagnosis, repair, calibration, and field replacement. |
| Performance | Ability of the system to provide accurate data from sensors to within 10%. |
| Area coverage | Ability of voice and data radio transmissions to cover 90% of remote stations 90% of the time, to and from the central facility. |
| Availability | Ability to meet a system availability figure of 0.98. |
| Cost | Ability of the systems life cycle cost to be less than or equal to a prescribed amount. |
| IOC date | Ability to become operational on or before a specified initial operational capability (IOC) date. |

both maintainability and performance. That is, should a design feature that provides maintainability and performance characteristics require that a lessor number of remote units meet coverage requirements, the coverage requirements will be subordinated.

Placing system availability fourth on the priority list means that the customer will tolerate some outages, but, while the system is up, performance and coverage take precedence. In all systems, system availability must be very carefully defined as early as possible. There are a number of ways to accomplish this, which are discussed in Appendix B.

The customer is willing to spend money to achieve maintainability, performance, coverage, and availability. Note that, if cost is listed as the highest priority, the entire activity is referred to as design-to-cost exercise. Note also that, in the present example, the complete life cycle cost is specified. In some instances, only development costs are included on the priority list.

Finally, the customer will tolerate some delays in system delivery in order to meet requirements imposed by the other characteristics.

Clearly, none of these characteristics are independent of any of the others. The maintainability and availability issues are intimately related, since availability includes consideration of the MTTR and cost is always a design issue throughout. That is why the word "competing" appears in the term prioritized competing design characteristics.

Further, the use of PCDCs does not imply unlimited prioritization. The fact that cost is listed fifth does not mean that funding is unlimited. Rather, it means that the customer is not strictly designing to cost and that some flexibility with respect to expenditures exists. Nor is the delivery date totally open ended; rather, the customer may tolerate extending the IOC by weeks or possibly even months if it can be shown that a clear advantage in meeting higher priorities can be realized.

It is not necessary to pin down the exact limits of tolerance in advance. The listing is meant to be more of an agreed-upon guide for design and design trade-off activity. When any requirement is in jeopardy, its impact must always be discussed with the customer.

The PCDC concept is also of value in avoiding surprises and potential disappointment on the part of the user community. Consider, for example, a deep space probe system whose major purpose is to perform scientific exploration at a distant planet. The spacecraft will be designed to carry a number of instruments. But, surprisingly, the number of instruments to be carried—that is, the science gathering capability of the platform itself—may be of the lowest priority. Consider that the target planet inherently imposes a specific launch window for a given launch vehicle, propulsion system, and mission trajectory design. If the launch window is missed, there will be no mission—hence, no science. Further, if the size and weight of the probe exceed certain limits, the launch vehicle will be inadequate—hence, no launch. If the communications subsystem consisting of transmitters, receivers, and antennas cannot provide sufficient effective radiated power to reliably send the science data back, then the system fails. We thus have the interesting situation that meeting a launch date with stringent specifications on mass, size, and communications capability, to name only a few factors, takes precedence over the number of instruments on the spacecraft, even though the purpose of the entire enterprise is to accomplish science. Thus, science becomes one of the lower priorities, and it is entirely possible that the final number of instruments may be reduced on the spacecraft over that which was originally intended in order that the mission be achieved at all. In such systems, meeting a launch date is typically the highest priority. Consequently, if a science instrument is not completed on time or if unforseen design trade-offs use up its allocated space or mass, it may simply be dropped and a lessor number of instruments may be flown. It is easy to understand why a clear grasp of such priorities must be well understood in advance by all project and investigating science personnel.

By way of further example of the insights that can result from a thoughtful consideration of priorities, consider the PCDCs for a rapidly deployable mobile gun system (RDMGS). The purpose of the system is to allow for rapid deployment anywhere in the world of a highly mobile weapon capable of attriting the most formidable of adversarial tanks. In this example, rapid deployability translates into a requirement to be shippable by any and all means of existing transportation, including the ability to be lifted by a helicopter. Lethality requirements for the specific threat to be engaged suggest that at least a 75-mm to 105-mm primary weapon be mounted on the RDMGS.

One hard reality of these requirements is that a small tank capable of being lifted by a helicopter would not be capable of carrying a significant amount of weight devoted to protective armor. A main battle tank typically weighs some 50 tons, while an RDMGS-type vehicle could not weigh much more than 20 tons total. Thus, the survivability of such a vehicle must clearly depend on its lethality and mobility (i.e., its ability to acquire, identify, and attack quickly and accurately and then move), as opposed to its protective armament.

It is further assumed in this example that, while establishing user needs, discussions indicated that the military senses a pressing need for such a vehicle; thus the IOC date is firmly established.

It is also clear that logistics support requirements for the RDMGS will be quite different from those associated with the more conventional use of a tank-type fighting vehicle. The RDMGS will be substantially separated from any tangible support mechanism while executing a given mission, but each mission is also likely to be quite short. These factors have an impact on system availability design trade-offs.

With these considerations in mind, the set of PCDCs proposed for the RDMGS is shown in Table 12-2.

The following rationale is suggested for the RDMGS PCDCs.

Transportability is ranked as the highest priority, since the inability of the RDMGS to be rapidly deployed violates the basic mission concept. Succinctly—if it can't get there, it can't do anything.

Once the vehicle is successfully deployed, the major purpose becomes to defeat the threat in order to accomplish the mission. Thus, lethality is ranked second.

Given its lethality, the field commander must be able to move his or her assets quickly to desired positions over specified obstacles. Mobility is deemed to be an important part of the hit and move concept, but it is subordinate to lethality and transportability.

The ability to field this asset in a useful time frame is considered the next most important priority.

Because of the short duration of proposed missions, coupled with the hard fact that the RDMGS will be a vulnerable entity, vehicle sustainability is rated next.

Cost, always an important factor, is rated as subordinate, within limits, to all of the higher priorities.

**TABLE 12-2 Prioritized Competing Design Characteristics for the RDMGS**

| | |
|---|---|
| Transportability | Ability of the RDMGS to meet deployment requirements by fixed-wing aircraft, helicopters, road, rail, and amphibious means. |
| Lethality | Ability of the RDMGS to identify, engage, and defeat specified threat targets. |
| Mobility | Ability of the RDMGS to meet specified maneuver requirements on land and water. |
| IOC date | Ability for the RDMGS to meet its initial operational capability date. |
| Sustainability | Ability of the RDMGS to meet requirements for reliability, availability, and maintainability. |
| Cost | Ability to field the RDMGS within a life cycle cost limit. |
| Survivability | Ability of the RDMGS to avoid and/or withstand threat. |

Due to the inherent light weight of the RDMGS and the rapid evolution of advances in anti-armor weaponry, it is felt that excessive expenditures of money and design activities to achieve a high degree of survivability for the RDMGS would have a low probability of paying off. From the mission perspective, the burden of survivability shall be carried indirectly by the higher-priority design characteristics of lethality, mobility, and the ability of the RDMGS to be removed rapidly from the chosen area of engagement upon termination of the mission. It is also clear that, with existing technologies, no amount of money can purchase sufficient protection for a vehicle of the required weight in the required mission setting.

Clearly, the RDMGS platform is to be designed for a dangerous and highly aggressive mission. After careful and stern objective consideration, system design priorities have emerged that place the lives of the combatants themselves at the lowest of priorities—a condition that, mentioned out of context, would not likely be acceptable. Military assets are not generally designed with such an outcome. In this case, the realization of priorities emphasizes that the survivability issue is clearly relegated to a need for a high degree of training, operational coordination, and use of the element of surprise. The example is a realistic one and emphasizes the importance of thoroughly considering priorities early in the systems engineering process.

Detailed priorities for consumer products are quite different from the MPGS example. Principles of the modern quality movement, for instance, dictate that the highest priority for consumer products is customer satisfaction and that among the lowest of priorities is development cost. This tenet is rapidly replacing older management by objective concepts, such as "decrease production costs by 10%" or "increase sales by 5%," which are internally directed priorities for product development. For consumer products, the term "customer satisfaction" typically translates into performance, which in turn maps onto the meeting of functional requirements, including maintainability. In any case, it is not sufficient to simply state that quality is a number one priority. The goal should be to further delineate a set of specific, well-defined priorities, such as performance, availability, delivery date, operability, cost, and so on.

Priorities can never be taken for granted and, unless clearly stated, will constantly differ among all participants throughout the design and implementation process. It is very important to establish and agree upon a set of PCDCs in the early going, if for no other reason than to avoid unnecessary and extensive discussions and confusion as the implementation unfolds. As the examples have shown, it can also be a very enlightening process in causing one to consider the consequences of meeting top-level requirements in a structured manner that brings factors to the fore that would surely be less defined without going through the exercise.

## HOW TO USE PCDCs

This section provides an example of a design trade-off analysis methodology. The methodology is generic. The example will furnish a structured approach to problem definition and will pinpoint specific areas where further action items should be assigned. The value of the methodology lies in the fact that it is basically invariant from system to system and leads to the logical definition of pertinent issues to be studied at the right level.

The design trade-off selected for this example is the use of a 105-mm versus a 75-mm primary weapon for the RDMGS. The first step in the generic process is to construct a matrix of PCDCs and WBS level 3 items that may be affected by the design trade-off. Such a matrix for the RDMGS is shown in Figure 12-1. In this example, seven level 3 items are identified as being impacted.

The next step is to examine each intersection of the matrix and determine if there are any possible impacts caused by trade-off options on the level 3 item with respect to each PCDC. Sufficient time must be devoted to this process to insure completeness in considering all influences and related aspects. For each potential impact identified, a number is noted in the appropriate intersection. These numbers will refer to notes that further define the nature of each impact. Figure 12-1 shows that there are six distinct issues that need to be addressed. It is assumed in this example that the IOC

| | Affected Level 3 Items | | | | | | |
|---|---|---|---|---|---|---|---|
| | Hull/ Frame | Suspension | Power Package | Drive Train | Aux Sys. | Turret Assem. | Armament |
| PCDC's | | | | | | | |
| Transport-ability | (1) | (1) | (1) | (1) | (1) | (1) | (1) |
| Lethality | | | | | (2) | | (3) |
| Mobility | | | (4) | | | | |
| IOC Date | | | | | | | |
| Sustain-ability | | | | | | | |
| Cost | (5) | (5) | (5) | (5) | (5) | (5) | (5) |
| Surviv-ability | | | | | | (6) | (6) |

**FIGURE 12-1.** Matrix of design trade-off impacts.

date and RDMGS sustainability are not influenced by the 105-mm versus 75-mm choice.

The referred-to numbered notes entered in the figure are expanded upon next. For example:

1. Using the 105-mm as the primary gun will increase the weight of the armament and, potentially, of the turret assembly. There may also be more weight for the auxiliary system ammo storage area. This will certainly be true if an equal number of rounds is provided for each alternative. These increases in turn potentially impact the engine power, drive train, suspension, and hull/frame designs. Subsequent reductions in armament may be required to meet transportability requirements.
2. It is likely that a smaller number of rounds can be accommodated by the 105-mm auxiliary system configuration, which decreases the effective mission length. The lethality issue relates to a decrease in effective time on station versus the higher probability of defeating the threat with more formidable ordnance.
3. The higher probability of success of the 105-mm gun over the 75-mm gun may allow for a trade-off in armament.
4. Increased engine power may be required to maintain mobility requirements due to an increase in weight for the 105-mm.
5. Cost differentials need to be assessed across the board.
6. The 105-mm turret assembly and primary armament will increase the RDMGS signature to some extent. Also, agility may be affected.

From this set of notes, specific information needs can be structured, with a reasonable probability that all factors have been considered. In this example, the following studies would be called for and initiated by the design team:

1. Determine total RDMGS weight impact. If transportability is compromised, study the extent to which armor reduction would be required. If the transportability requirement cannot be met or if survivability is degraded beyond acceptable limits, then the 105-mm option must be discarded.
2. Determine the differential in lethality capability for the total mission scenario as a function of:
   a. Rounds of higher lethality;
   b. Lesser number of rounds (mission length); and
   c. Firing rates.
3. Determine engine power and related enhancements to achieve mobility requirements.

4. Determine system cost differentials.
5. Determine impact of signature size on survivability.

As a reader, you are welcome to suggest other items that may be of interest in the trade-off example. The purpose of the example is to suggest a structure that focuses the consideration of impacts across all pertinent level 3 WBS items on each of the systems design priorities. It is a structure that guides the analyst in an effort to achieve completeness, and, at the same time, maintains the thought process at a consistent level.

The systems engineer is not expected to identify and resolve all the issues that are defined as a result of the trade-off analysis process. The SDT is typically made up of personnel with greater in-depth capabilities to support the required analysis. The systems engineer, however, should play a leading role in identifying potential trade-offs, and play the lead role in structuring the path of inquiry for each study undertaken. Considering level 3 items together with PCDCs provides a logical structure to achieve these goals.

## THE ROLE OF MODELING IN DESIGN TRADE-OFFS

Throughout system development, the systems engineer is constantly challenged to acquire and maintain visibility into the future. An important part of this understanding consists of gaining confidence that requirements and design concepts are realistic. Modeling should be used at any point in the systems engineering process when confidence in these issues is at all shaky. The purpose of modeling is to mimic the appearance or behavior of a system, subsystem, or assembly such that an improved understanding of unknown attributes can be acquired without the time and expense of building and testing the final product.

We are all familiar with concepts of early and rapid prototyping, breadboarding, mock-ups, wind tunnels, and so on, as well as operational research tools, such as simulation and analytic modeling. In one sense, these are all distinct disciplines and, indeed, whole careers have been devoted to the subtleties of their expert pursuit. From the systems engineering viewpoint, however, the use of each of these tools has a common motivation—to reduce uncertainty through the study of physical and/or mathematical behavioral likenesses.

The need to reduce uncertainty can arise at any point in the systems engineering process, from determining requirements to validating the producibility of the final baseline product. Each technique exhibits its own value, with respect to the degree to which it can reduce uncertainty as a function of the specific issue at hand. The systems engineer should have a clear perspective as to the purpose for undertaking a particular modeling exercise.

There are two fundamental purposes for undertaking modeling. The first is to assist in defining requirements themselves, and the second is to verify an implementation concept designed to meet previously defined (and hence controlled) requirements. This may seem like a moot difference, but remember that the systems engineer is subject to criticism when requirements "change," particularly as the configuration management structure becomes more formal and change becomes more difficult. It is not unusual for a modeling exercise in support of a design process to turn up surprises that then result in the need to alter requirements. It is often advantageous to conduct modeling early in the process to verify that specific demanding requirements can be met. If there are any doubts, it is best to recognize this situation and schedule the work at the requirements definition phase to reduce the risk of "changes" later in the process. The choice of modeling techniques and the timing of their use is inherently related to the systems engineering paradigm in use on a given project.

It is also important that management understands the strategies employed and their rationale. For example, early prototyping is commonly used to understand user requirements. In this application, the modeling work is not intrinsically a top-down detailed design process. Rather, the prototype is brought together quickly, with emphasis on resolving specific issues. With luck, much of the prototype development may be usable in the actual detailed design that is to come later, but this is not always the case—nor should it even be counted on. Management often misunderstands this, asking, "If you have built a prototype that works, why do you need further design time?" Thus, it is important to understand and to make known the specific goals and expected products of modeling at any particular stage. Such steps to reduce uncertainty should be a planned and well-advertised part of all schedules.

Table 12-3 provides a breakdown of approaches to modeling by class and type, with examples. Both the static and dynamic classes of models are divided into physical and mathematical types. The following paragraphs discuss the roles of these modeling tools in the synthesis of requirements and design, and provide general guidelines for timing and conditions for their use from the systems engineering viewpoint.

Static models are those primarily concerned with steady state conditions. Examples of static physical models include mock-ups, (such as cockpit, console, or work station layouts) and the physical laying out of simulated equipments, cables, and so on, to mimic design and/or placement concepts. Mock-ups are generally used to bring a more precise definition of requirements affecting human factors, safety, and physical fitness issues. Static loading models on structures such as airplane wings are employed to verify

**TABLE 12-3 Roles of Modeling in the Synthesis of Requirements and Design**

| Class | Type | Examples |
|---|---|---|
| Static | Physical | Mock-ups, positioning, static loading, etc. |
| | Math | Queueing, reliability, steady state analysis, etc. |
| Dynamic | Physical | Breadboards, brassboards, early prototypes, production prototyping, etc. |
| | | Analogues—wind tunnels, ship hull models, biological models, etc. |
| | Math | Simulation—GPSS, SIMSCRIPT, etc. |
| | | Analog—analytic equations |

the ability of a design concept and specific materials to meet previously defined requirements.

Static mathematical models include such tools as queueing analysis, availability, reliability, and maintainability analysis and any of a large class of analytic representations that predict steady state conditions.

The disciplines of queueing analysis and system availability analysis are of extensive value to the systems engineer. The majority of complex systems built today are characterized by the use of computers, contention for resources, and stringent system availability requirements. In such systems, queueing analysis (or simulation, when required) and availability analysis are the principal tools used to initially determine top-level system architectures. A specific application example is given later in this chapter.

Dynamic physical models include the construction of crude approximations, near replicas, use of scaled-down models, and experimentation with systems that exhibit similar physical characteristics.

Tables 12-4 through 12-7 summarize principal properties of breadboards, brassboards, early prototypes, and prototypes. Again, these terms are familiar to most engineers, yet it is useful to characterize them with respect to the timing and rationale for their use in support of the systems engineering process.

Breadboards provide a quick and dirty look at concept verification. Electronic breadboards are usually a stringy mess and do not provide a good

**TABLE 12-4 Design Approach—Breadboards**

| | |
|---|---|
| *Characterized by:* | A stingy mess |
| *Confidence level:* | low |
| *Rationale:* | See if Charlie's crazy idea really works |
| *Be sure to:* | Keep a good engineering notebook |

**TABLE 12-5 Design Approach—Brassboards**

| | |
|---|---|
| *Characterized by:* | A tidy semi-kludge |
| *Confidence level:* | Medium |
| *Rationale:* | Test out a few unknowns before committing |
| *Be sure to:* | Keep a good engineering notebook and thoroughly test |

**TABLE 12-6 Design Approach—Early Prototype**

| | |
|---|---|
| *Characterized by:* | Near replica of working environment |
| *Confidence level:* | Low to medium |
| *Rationale:* | Definition of unknown requirements |
| *Be sure to:* | Keep a good engineering notebook and minimize configuration control |

**TABLE 12-7 Design Approach—Prototypes**

| | |
|---|---|
| *Characterized by:* | Replica of serial number one |
| *Confidence level:* | High |
| *Rationale:* | Build and test serial number one before production |
| *Be sure to:* | Duplicate production model and thoroughly test |

environment for tolerance testing of factors such as the effects of capacitance, cross-talk, and so on in a finished version. Breadboards provide an initial test bed only. When they work, a more tidy representation that is closer to the final product should always follow.

Brassboards generally represent a more serious attempt to mimic the behavior of the final product, particularly in terms of component placing, connector lengths, and packaging. Portions of brassboards that involve high-risk areas may be constructed as near replicas of the envisioned final product and may be of sufficient integrity to undergo considerable testing.

In this classification, a distinction is made between the terms "prototype" and "early prototype." Prototypes are virtual replicas of serial number one and are built primarily for extensive testing and for insuring producibility.

Breadboards, brassboards, and prototypes are typically built to support design trade-off studies or to verify the capability of more mature designs to meet previously established requirements.

Early prototypes are quite different in that they are usually used to help define requirements or, in some instances, trade-off analyses. Their design is characterized by providing a near replica of user interaction, even through the location might be quite different. A further principle characteristic is that the development environment is ideally capable of rapidly accommodating potential design changes in response to user feedback.

Early, or rapid, prototyping finds its greatest value when the user is unable to communicate exactly what his or her requirements are, but is able to recognize, select, or provide guidance as to what design features are desired when confronted with appropriate alternatives. Such conditions commonly exist, for example, when building the user interface for a new military command and control system, for a novel management information system, or for supporting market research for consumer products.

It is not always the case, however, that rapid or early prototyping can be of value when requirements are unknown or fuzzy. In some cases, exact requirements cannot be articulated because no one inherently knows what they are, nor can they recognize them in any detail when shown. This is the very real case, for example, with the space station whose mission will extend beyond a time frame that enables anyone to foresee its exact usage. In cases of this sort, the systems engineering process should be structured around the rapid development concepts presented in Chapter 2.

Examples of physical analogues include constructing scale models or experimenting with systems that exhibit behavioral characteristics similar to the system of interest. Familiar scale models include ship hulls in water tanks and aircraft models in wind tunnels, automobile designs, and so forth. Biological models are used as analogues of human systems (e.g., the hind leg of a cat represents the closest analogue to the musculature of the human lower extremities).

Dynamic mathematical models include transaction-oriented simulation or discrete, simulation tools, such as SIMSCRIPT and GPSS, as well as the use of differential and integral calculus analogs implemented on digital and, less frequently, analog computers. Typical uses in these arenas include analyzing and designing computer systems, communications systems, fire control systems, engine performance, marketing systems, physiological systems—virtually any continuous system. All continuous systems are basically the same.

The classifications and examples presented here are, of course, not exhaustive. Rather, the structure presented is intended as a conceptual framework that can easily be amended.

For the systems engineer, the choice of model implementation is predominantly driven by the level of confidence associated with the inquiry at hand—that is, the depth of inquiry required and an understanding of whether the issue has to do with establishing requirements or with design verification.

Model structures are often improperly chosen. Engineers who spend their lives in rapid prototyping laboratories will naturally tend to favor rapid prototyping to resolve as many issues as possible. I have also seen simulation experts expend a great deal of funds and time to resolve system performance issues that could have been determined in one-tenth to one-hundredth the time through the proper application of simple queueing theory.

In summary, it is best to clearly consider the following points when electing to employ modeling techniques:

1. Discuss the need and strategy for modeling with the SDT. Reach a consensus. Derive and write down a succinct statement addressing the specific question to be answered in measurable terms. The goal must be limited and well-defined. One or two sentences for inclusion in the minutes should be sufficient.
2. Construct a realistic schedule to achieve the acquisition of resources, execution of the analysis, and required iterations. The schedule must support the project schedule.
3. Assign the lead responsibility for resolution of the question to a single individual.
4. Be sure that all required data input and/or drivers for the model can be accurately acquired in a timely manner. These are often overlooked.
5. Limit the activity to addressing the question at hand.
6. Inform management of your plans.

## Determining System Architecture—An Example

All viable system architectures must meet performance and availability requirements at a minimum. The ability of a proposed architecture to meet performance requirements can be estimated through use of appropriate modeling techniques. The ability of a proposed architecture to meet availability requirements can be estimated through use of RAM analysis. The systems engineer should use both of these valuable techniques to bring increasing detail to estimates of performance and availability, beginning at the point of establishing early realistic functional requirements and continuing to the point where actual measurements can be made to validate these system characteristics. This section draws on material in Appendices A and B to provide an example of a typical top-level architectural study on a computer-based system, using an analytic queueing model followed by RAM analysis.

Queueing theory can be used to rapidly gain a fundamental understanding of weak links in the architectural chain of virtually all computer-based systems. This is true even though simulation may eventually be required for a deeper understanding of all aspects of system behavior.

Consider a computer-based system that consists of the four units shown in Figure 12-2, Part 1. The units might consist of CPUs, tape units, disks, communications controllers, and so on. Each unit has a mean service time for

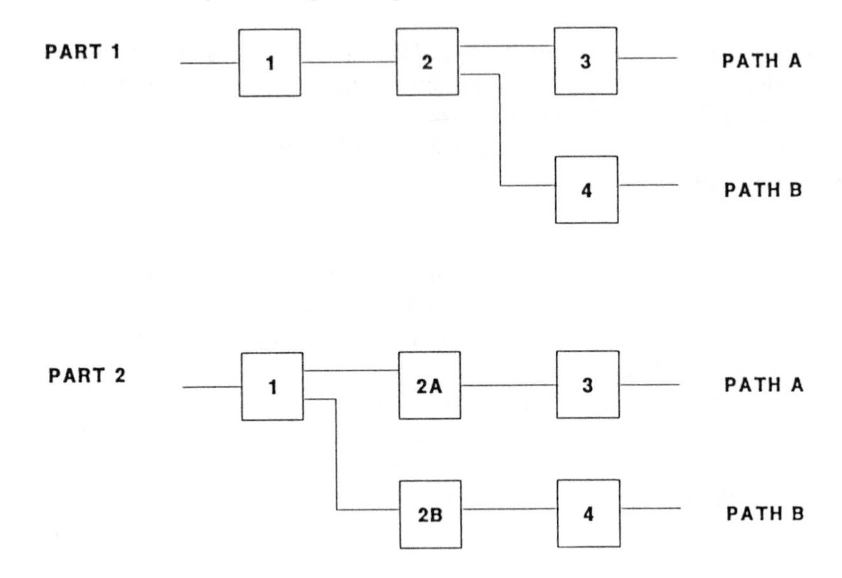


**FIGURE 12-2.** Performance driven architecture example.

the message mix, and the system experiences a peak arrival rate load for requests for service.

Table 12-8 lists the mean service times and transaction arrival rates for each unit to be used in the example.

Note that unit 1 receives 8 requests for service per second and generates the same load for unit 2. Unit 2 also generates a transaction load of 8 per second, which is then evenly divided between units 3 and 4, each receiving 4 requests for service per second. The mean service times for units 1, 2, 3, and 4 are given as 70, 100, 50, and 100 milliseconds, respectively.

We shall also assume in this example that the system functional requirement calls for a total average queue time, or average response time, of 500

**TABLE 12-8 Mean Service Times and Arrival Rates for Architecture Example**

| Unit | Mean Service Time in Seconds | Arrival Rate per Second |
|---|---|---|
| 1 | .070 | 8 |
| 2 | .100 | 8 |
| 3 | .050 | 4 |
| 4 | .100 | 4 |

milliseconds for Path A and 600 milliseconds on Path B. At this point in the design, you have stipulated a desire to maintain a 10-percent margin on response times (see Chapter 15 for margin management guidelines). Assume that the system availability requirement for Path A is .9800 and the system availability requirement for Path B is .9900.

Finally, as the process of determining an appropriate system architecture unfolds, the systems engineer will need to make decisions with regard to the feasibility of trade-offs. Such decisions should be routinely guided by the previously established PCDCs.

For the present example, we shall assume the PCDCs in Table 12-9 have been formulated by the SDT.

Table 12-10 presents the results of calculating unit utilizations [E(ts) * E(a)], the resulting mean waiting times E(tw) for each unit, the resulting mean time in queue E(tq) for each unit, and the total system queue time E(tqs) for both Paths A and B.

It is evident that the response time requirements for Paths A and B are not met. It is also clear that the major contributor to the excessive values for E(tqs) is the behavior of unit 2 and, in particular, the mean waiting time E(tw) of some 400 milliseconds. Note that, even if our estimates that went into the queueing calculations were off by as much as 10 percent, it is still clear that

**TABLE 12-9 Prioritized Competing Design Characteristics for Architecture Example**

1. Adherence to schedule
2. System availability
3. System performance
4. System operability
5. Cost

**TABLE 12-10 Queueing Results for Part 1 *Architecture Example***

| Path | Unit | Utilization | E(tw) | E(tq) | E(tqs) |
|---|---|---|---|---|---|
| A | 1 | .56 | .089 | .159 | |
| | 2 | .80 | .400 | .500 | |
| | 3 | .20 | .013 | .113 | |
| | | | | | .772 |
| B | 1 | .56 | .089 | .159 | |
| | 2 | .80 | .400 | .500 | |
| | 4 | .40 | .067 | .167 | |
| | | | | | .826 |

unit 2 is causing the greatest relative delay by a factor of at least four. The waiting time is excessive on unit 2 because of the high utilization of .8. This is representative of the kinds of insights that can rapidly be gained through the use of queueing models.

There are two basic ways to reduce the utilization on unit 2. The first is to reduce the mean service time, and the second is to reduce the arrival rate. The mean service time can be reduced by obtaining faster hardware or by effecting programming efficiencies. The arrival rate can be reduced by increasing the number of units by using multiple single-servers or by using a single multiple-server.

As systems engineer, you will want to explore both of these. One beginning point is to ask the systems software engineer (SSE) on your design team to lead a study to determine the feasibility and effort involved in reducing the mean service time. Working the problem backwards reveals that the mean service time for unit 2 will have to be reduced by almost a factor of one-half, say 50 milliseconds, in order to achieve a utilization near .40, which would allow for the required system delays. In the meantime, you begin to look at costs for duplicating unit 2 for a dual single server configuration.

Within a week, the SSE reports to the design team that the best improvement that can be realized in the mean service time for unit 2 would be a 20-percent reduction, from 100 milliseconds to approximately 80 milliseconds. The SSE reports that this can be achieved through a combination of adding a math co-processor and improving application programming efficiencies. Improvements beyond this level would require substantial technical inquiry involving schedule impacts. The CPU itself cannot be upgraded any further.

Table 12-11 shows results of applying the queueing model again, this time with a mean service time of 80 milliseconds for unit 2. Path B now shows an

**TABLE 12-11 Queueing Results for Part 1 with Unit 2 E(ts) = 80 Milliseconds**
***Architecture Example***

| Path | Unit | Utilization | E(tw) | E(tq) | E(tqs) |
|---|---|---|---|---|---|
| A | 1 | .56 | .089 | .159 | |
| | 2 | .64 | .142 | .242 | |
| | 3 | .20 | .013 | .113 | |
| | | | | | .514 |
| B | 1 | .56 | .089 | .159 | |
| | 2 | .64 | .142 | .222 | |
| | 4 | .40 | .067 | .167 | |
| | | | | | .548 |

estimated response time of 548 milliseconds, which meets the 600 millisecond response time requirement. The margin, however, is only about 9 percent, which is slightly below the desired 10-percent margin. Path A, with a newly estimated response time of 514 milliseconds does not meet the 500 millisecond response time requirement for that path.

The feasibility of reducing the utilization of unit 2 through halving the arrival rate by replacing unit 2 with two units, 2A and 2B, is next considered. This configuration is depicted in Part 2 of Figure 12-2. The 8 transactions per second generated by unit 1 are now evenly divided between units 2A and 2B, such that the new arrival rate of 4 transactions per second is received at each. This should halve the utilization from .8 for unit 2 to .4 for each of the units 2A and 2B. The queueing analysis now results in the values shown in Table 12-12.

The model estimates an average response time for Path A of 439 milliseconds, which meets the 500 millisecond requirement with a margin of 14 percent. The Path B estimated response time of 493 millisecond also meets its 600 millisecond requirement with a margin of 21 percent. We also note that the unit utilizations across the system are now more balanced. It is also evident that, should the traffic load increase on the system, the next potential response time problem will occur at unit 1, whose estimated utilization is higher than that of units 2, 3, and 4.

We have found at least one solution. The question that now remains is whether it is the more feasible solution. It may still be possible to investigate methods of reducing the mean service time at unit 2, or perhaps a combination of improvements between units 1 and 2. The method for this analysis is merely a variation on what has been carried out above. In our example, however, the SSE has reported that there would be considerable cost and

**TABLE 12-12 Queueing Results for Part 2 with Units 2A and 2B Replacing Unit 2**
***Architecture Example***

| Path | Unit | Utilization | E(tw) | E(tq) | E(tqs) |
|---|---|---|---|---|---|
| A | 1 | .56 | .089 | .159 | |
| | 2 | .40 | .067 | .167 | |
| | 3 | .20 | .013 | .113 | |
| | | | | | .439 |
| B | 1 | .56 | .089 | .159 | |
| | 2 | .40 | .067 | .167 | |
| | 4 | .40 | .067 | .167 | |
| | | | | | .493 |

schedule impacts in pushing the technology for further mean service time reductions.

It is interesting to note that it is generally easier to reduce utilizations by duplicating functions (i.e., reducing arrival rates by at least one-half) than by reducing service times. While some service time reductions may initially be relatively simple, service time reduction typically becomes exponentially more difficult beyond a certain point. On the other hand, duplicating unit 2 in the system also costs money.

At this point, the systems engineer turns to the previously determined PCDCs for guidance. These are given in Table 12-9 for our sample project. The project is schedule driven, which argues against further efforts to study and implement service time reductions. System performance, which includes the ability to meet response time requirements, has a higher priority than cost. That is, the customer is quite willing to spend money to meet performance requirements and most certainly desires to maintain the schedule. System operability is not impacted either way by the particular design decision at hand. Thus, the conclusion is that the additional cost of splitting unit 2 into two units, 2A and 2B, is most consistent with the project PCDCs.

Having developed a first pass at a system architecture based on performance, we now turn to examining the system availability characteristics of the configuration. Figure 12-3 presents the availability models for Paths A and B. We begin by considering Paths A and B without redundancy. Each path includes power and environmental control equipment in series with

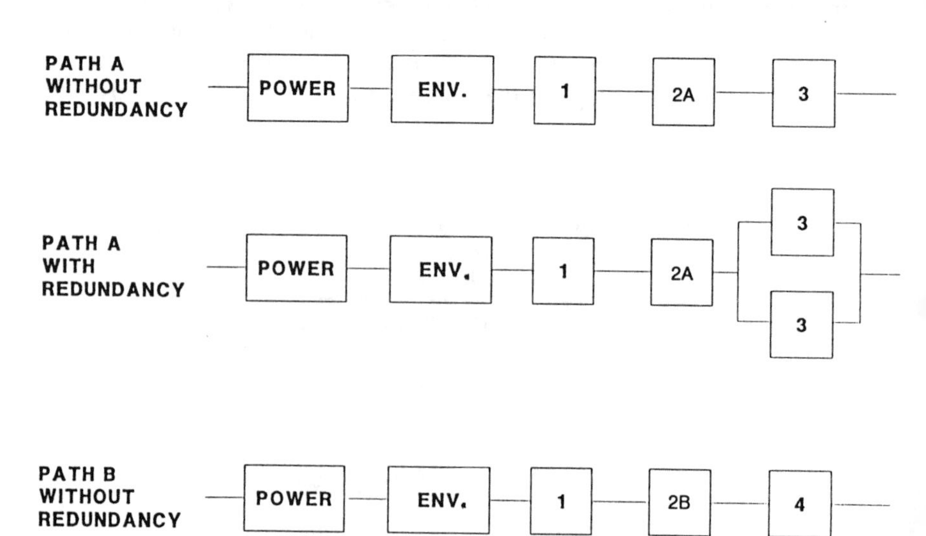


**FIGURE 12-3.** Availability driven architecture example.

their respective units—units 1, 2A, and 3 for Path A and units 1, 2B, and 4 for Path B. The assumed MTTF and MDT values for all units involved are given in Table 12-13.

The system availability for the complete serial path is then determined, using either Equation 21 or 22 in Appendix B. The results of these estimates are listed under the first column, headed Case A1, in Table 12-14.

The calculated availability value of .9731 does not meet the functional requirement of .9800. The unit with the least availability is unit 3, with a value of .9850. Investigating the use of simple redundancy for unit 3 yields the newly calculated value for Path A availability shown in Table 12-14 under Case A2. The equivalent MTBF has improved by a factor of 1.5 (see equation B-15 in Appendix B), but this improvement still yields an insufficient value for path

**TABLE 12-13 MTBF and MDT Values for Architecture Example**

| Unit | MTBF (hrs) | MDT(hrs) |
|---|---|---|
| Power | 4,380 | 18 |
| Environmental control | 8,000 | 8 |
| 1 | 20,000 | 18 |
| 2 | 5,000 | 36 |
| 3 | 2,500 | 36 |
| 4 | 7,000 | 6 |

**TABLE 12-14 Path A Availability Calculations for Three Cases**

| | Availabilities | | |
|---|---|---|---|
| Unit | Case A1 | Case A2 | Case A3 |
| Power | .9959 | .9959 | .9959 |
| Environmental control | .9990 | .9990 | .9990 |
| 1 | .9991 | .9991 | .9991 |
| 2 | .9929 | .9929 | .9929 |
| 3 | .9850 | .9905 | .9996 |
| Path A availability | .9731 | .9777 | .9865 |

Case A1—No redundancy
Case A2—Simple redundancy of Unit 3
Case A3—Unit 3 redundant with immediate replacement

availability of .9777. We next consider redundancy of unit 3 with immediate repair (equation B-16 of Appendix B). The result of this calculation is shown as Case A3 in the table. The predicted path availability of .9865 now meets the requirement of .9800.

Having found one possible solution, we now turn to Path B. As with Path A, the availability of each unit in Path B is first calculated and then the availability for the entire serial path is determined. Results of these calculations are shown under Case B1 in Table 12-15. The resulting availability of .9861 does not meet the requirement of .99 for Path B.

The two units in Path B exhibiting the least, or weakest, unit availabilities are the power unit and unit 2. These units represent the most logical candidates for modification. Review of Table 12-13 suggests that reducing the 36-hour MDT for unit 2 presents one approach. The 36-hour MDT is based on the vendor's response specified in the maintenance contract. Another approach is to increase the MTBF of the power configuration beyond that which the existing power grid can provide through use of a uninterruptable power supply (UPS).

The design team is tasked through a specific action item to identify support mechanism scenarios that would substantially reduce the MDT for unit 2. The exact value for MDT required for unit 2 to bring the availability of Path B to .9900 is calculated using the availability equations as 16 hours.

The availability equations are used again to estimate the availability for Path B using an UPS. The UPS MTBF is estimated at 500,000 hours, which should effectively remove the power component from adversely affecting the

**TABLE 12-15 Path B Availability Calculations for Two Cases**

| | Availabilities | |
|---|---|---|
| Unit | Case B1 | Case B2 |
| Power | .9959 | .9999 |
| Environmental Control | .9990 | .9990 |
| 1 | .9991 | .9991 |
| 2 | .9929 | .9929 |
| 4 | .9991 | .9991 |
| Path B availability | .9861 | .9900 |

Case B1—Normal grid power
Case B2—Grid power with UPS

path availability. The results of this calculation are shown under Case B2 in Table 12-13, which provides a Path B availability of .9900. A second action item is given to determine costs of an UPS, which would also allow the availability requirement of .9900 to be met.

In less than 2 weeks, the team members respond to the action items. It is reported that the MDT for unit 2 can be reduced to 16 hours if specific spare parts for unit 2 are kept in inventory and an additional crew member with vendor training is added to the maintenance staff. It is also determined that the one-time cost of procuring an UPS is less than the life cycle cost of adding an additional, highly trained staff member.

Armed with this information, the systems engineer then turns to the project PCDCs once again for guidance. Neither alternative in question affects schedule, performance, or operability of the system. This leaves cost as the remaining top-level decision criteria, despite the fact that cost has the lowest priority among the PCDCs.

Since the life cycle cost for lowering the MDT for unit 2 is greater than the one-time cost for the UPS, the decision is made to include an UPS in the system. The system architecture now appears, as shown in Figure 12-4.

This example is by no means been exhaustive. There are many other specific trade-offs and combinations that might have been considered. The example, however, has served to clarify the process of interaction between modeling, RAM analysis, PCDCs, and the roles of the systems engineer and of the SDT in arriving at a top-level system architecture. Clearly, there are many architectural alternatives to consider in meeting performance and availability requirements, including transaction load reduction, mean service time reduction, the use of redundancy, the use of higher-reliability parts, and the complete logistics support of mean downtime. This section has provided a methodic guideline for that process.

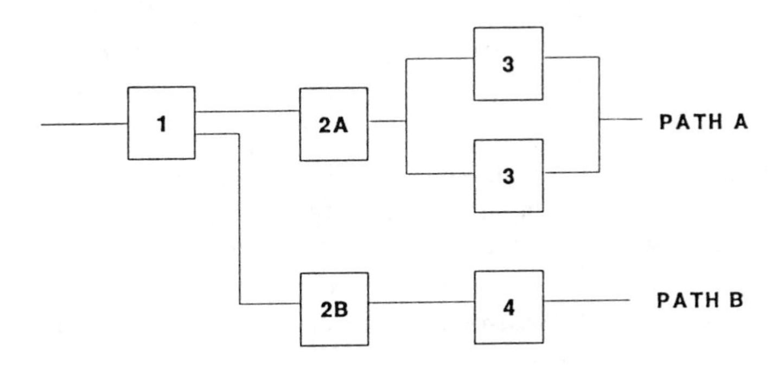


**FIGURE 12-4.** Final architecture—architecture example.

## RISK ANALYSIS

Among the many responsibilities of the systems engineer is to identify and quantify technical risk and the formulation of alternative backup strategies when required. The use of PCDCs facilitates a consistent and structured approach to the technical evaluation of trade-offs. It is can also be useful, when appropriate, to bring quantization to the level of risk associated with alternative approaches.

One common method of quantifying risk involves making estimates for the *probability of failure, P(f)*, and the *consequence of failure, C(f)*. These quantities are then used to calculate an overall *risk factor, RF*, which is a measure of the likelihood of failure [1].

The value $P(f)$ is a measure of the lack of success of a given system, subsystem, or lesser part of a system. Typical considerations are the maturity and complexity of hardware and software. The approach also allows for inclusion of other factors. The equation for $P(f)$ may be expressed as

$$P(f) = \frac{P_{MH} + P_{MS} + P_{CH} + P_{CS} + P_D}{5} \qquad (12\text{-}1)$$

where:

$P_{MH}$ = Probability of failure due to hardware maturity;
$P_{MS}$ = Probability of failure due to software maturity;
$P_{CH}$ = Probability of failure due to hardware complexity;
$P_{CS}$ = Probability of failure due to software complexity; and
$P_D$ = Probability of failure due to other factors.

The magnitude of the term in the denominator is adjusted to equal the number of terms in the numerator.

The value $C(f)$ is a measure of the lack of worth of an element because of its failure. Typical consequences of failure can impact technical, cost, and schedule factors. The equation for $C(f)$ may be expressed as

$$C(f) = \frac{C_T + C_C + C_S}{3} \qquad (12\text{-}2)$$

where:

$C_T$ = Consequence of failure due to technical factors;
$C_C$ = Consequence of failure due to cost factors; and
$C_S$ = Consequence of failure due to schedule factors.

The risk factor, *RF*, is the overall likelihood of failure and is expressed as

$$RF = 1 - [1 - P(f)] * [1 - C(f)] \qquad (12\text{-}3)$$

Table 12-16 lists sample rationale for assigning values for the components of probability of failure and consequence of failure for a system that will use existing computer hardware and software, with the exception of minor changes in displays. We assume that the system in question also involves research and development for a novel, high-performance transducer that is reduced in size and designed to replace an existing bulky technology. While the transducer concept is not mature, it is considered promising in that it is not, in itself, complex.

The values for $P(f)$ and $C(f)$ are calculated using equations 12-1 and 12-2 to yield

$$P(f) = .32$$

and

$$C(f) = .50$$

**TABLE 12-16 Sample Assignment and Rationale for Components of $P(f)$ and $C(f)$**

| Component | Value | Rationale |
|---|---|---|
| $P_{MH}$ | .8 | Transducer concept is new, so research is required |
| $P_{MS}$ | .3 | Existing software data processing package will be used, with minor display modifications |
| $P_{CH}$ | .2 | Computer hardware inherited, transducer concept involves low complexity |
| $P_{CS}$ | .2 | Low complexity in terms of number of software modules |
| $P_D$ | .1 | Negligible risk associated with vendors and subcontractors |
| $C_T$ | .7 | Significant degradation in performance if novel transducer concept fails and backup strategy to existing technology is required |
| $C_C$ | .5 | New transducer proof of concept expected early—moderate cost risks |
| $C_S$ | .3 | If new transducer concept not verified on schedule, revert to existing technology—minor schedule impact anticipated |

The overall risk factor, *RF*, is then calculated using equation (12-3) as

$$RF = 1 - (1 - .32)\,(1 - .50) = .66$$

Table 12-17 provides a guideline for the assessment of the risk factor *RF*. In the example given, the risk factor is considered to be moderate to significant.

While such rating factors can be of value in attempting to quantize the degree of anticipated risk, it should be noted that rating schemes are among the more simplistic approaches within the broader discipline of multi-attribute decision analysis. Further reading in the areas of risk management are suggested in the chapter references. [2], [3].

## SUMMARY

Trade-off studies are routinely conducted throughout the systems engineering process. While this chapter has focused on concepts and processes that have proven to be of value, there are, in fact, many ways to carry out such studies. Further, in some instances where such issues as national goals, strong user preferences, or politics dominate, it can be difficult to specifically quantize all rationale for decision making. Thus, issues of basic and significant import may be imposed on the system engineer and the SDT.

In the end, the systems engineer must make judgements, and there is no attribute of higher quality in these affairs than simple experience.

In this often difficult environment, perhaps there is no better instrument than the thoughtful definition and consistent use of a set of priorities in the form of PCDCs. These characteristics are of value in the early top-level stages of options analysis through to the highly specific trades that may be made in

**TABLE 12-17 Risk Factor Magnitude Guidelines**

| Magnitude | Interpretation |
|---|---|
| .1 | Negligible |
| .2 | Low |
| .3 | Minor |
| .4 | Minor to moderate |
| .5 | Moderate |
| .6 | Moderate to significant |
| .7 | Significant |
| .8 | High |
| .9 | Extreme |

the very final stages of development. PCDCs also set a tone of fundamental mission understanding that substantially contributes to team cohesion, efficient communication, and common purpose throughout the entire development effort.

**References**


1. *System Engineering Management Guide.* Lockheed Missiles & Space Company, Inc. 1983.
2. Keeney, Ralph, and Howard Raiffa. 1976. *Decisions with Multiple Objectives*. New York: John Wiley.
3. Keeney, Ralph, 1992. *Value-Focused Thinking*. Cambridge, Ma.: Harvard University Press.

# 13

# Fundamentals of Software Development Techniques

The purpose of this chapter is to enhance the ability of systems engineers without formal training in software development to communicate with software engineers. Systems engineers often have extensive backgrounds in software engineering. They rise to systems engineering positions through previous roles as software systems engineers. Those who do not have such strong backgrounds, typically hardware engineers, will naturally need to rely more heavily on the experience of the software systems engineer on their design team. In these cases, the need to communicate with regard to fundamentals of software development techniques is essential.

Hardware engineering has a well-entrenched lore related to applied mechanics technology, design, drafting, parts engineering, packaging, and fabrication. It is also easier to see hardware-related products unfold throughout the preliminary design, detailed design, and implementation phases of a project than it is to "see" software products. Thus, hardware milestones are somewhat easier to define and track. Further, because hardware design and implementation is so well established, the systems engineer can often rely on this expertise being in place in mature and well-run organizations.

The situation with regard to software systems engineering is not comparable. Through the years, many innovations have contributed to the evolution of software development techniques. From the development perspective, a central goal of these activities has been to devise methodologies that result in *consistent* product structures, independent of the individual developer. This condition is much more realizable in the world of hardware and remains an elusive target in the software environment.

While the structural and design concepts presented in this chapter represent meaningful advances toward consistency, there is still no "right" or "wrong" structure for a given problem. In the end, the ability of a program to meet user requirements and the ability of its architecture to withstand the trauma of operational maintenance is largely a matter of software development experience.

There is no asset so valuable to software development as that of experience. If you don't have it, it is then crucial that your software systems engineer bring years of successful experience to your team.

## STRUCTURAL CONCEPTS

This section provides an overview of basic structural concepts as they pertain to code, program modules, and the representation of data. Program structures that result from design methodologies are discussed in the next section.

### Code Level Concepts

In 1968, E. W. Dijkstra published a letter espousing the virtues of eliminating both forward and backward GO TO statements in coding [5]. The idea was not new, but it gained wide credibility with its publication. He contended that the GO TO's result is the de-structuring of programs and that they cause great havoc during program maintenance.

New code structures to replace the GO TO were soon widely accepted based on the concept that a fixed set of unambiguous control structure paradigms could be used to construct any program. Consistent use of these structures augments communication between programmers and users and facilitates software testing and maintenance. In particular, the constructs avoid the ambiguous and confusing use of sudden code stream exits.

Figure 13-1 shows a schematic representation of these basic paradigms. Program sequence is simply represented by a series of operations following the order in which they are carried out. The other paradigms provide for conditional operations. The IF-THEN-ELSE construct allows for execution of one of two alternative sequence paths, depending upon the outcome of a condition test at the decision diamond. Convention calls for the THEN path to be executed if the condition is true.

DO-WHILE and DO-UNTIL structures allow for looped operations. The DO-WHILE executes a condition test first and then executes a desired sequence as long as the condition is true. The DO-UNTIL first executes the desired sequences and then tests for the condition. When the condition is true, the DO-UNTIL loop is exited. The CASE construct allows for branching to one of a number of sequences when the appropriate case parameter is identified.

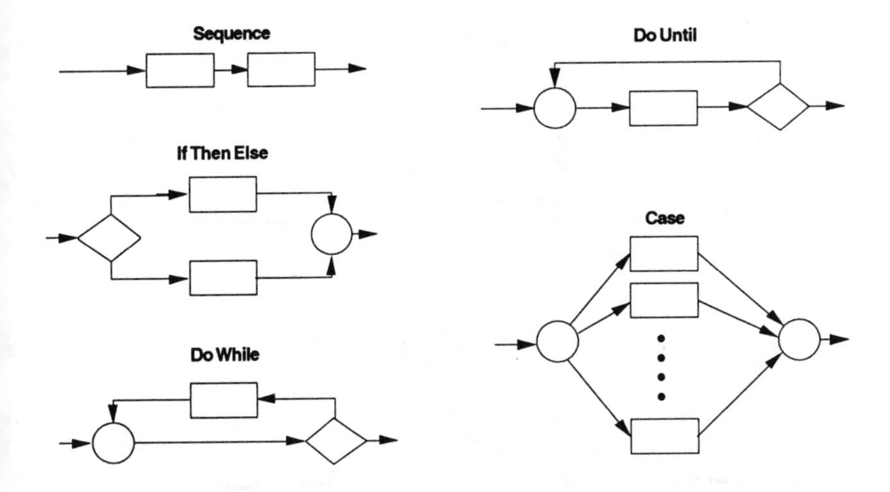


**FIGURE 13-1.** Control structure paradigms.

These structural concepts have remained with us since their introduction in the late 1960s and are considered fundamental to good coding practice.

## Module-level Concepts

At the software module level, there are a number of established structural considerations. G. D. Bergland provides a succinct description, [2]:

Modular programs can be characterized as

- implementing a single independent function,
- performing a single logical task,
- having a single entry and exit point,
- being separately testable, and
- being entirely constructed of modules.

Two major structural concepts that contribute to the realization of these characteristics are those of module *coupling* and module *cohesion.* Coupling and cohesion are designed to result in the qualities of functional independence and information hiding.

Module coupling is related to the amount of interaction between software modules. Figure 13-2 depicts both highly coupled and loosely coupled modules. Software is ideally designed such that there is a minimum of interaction between modules. This is usually attained by striving for functional indepen-

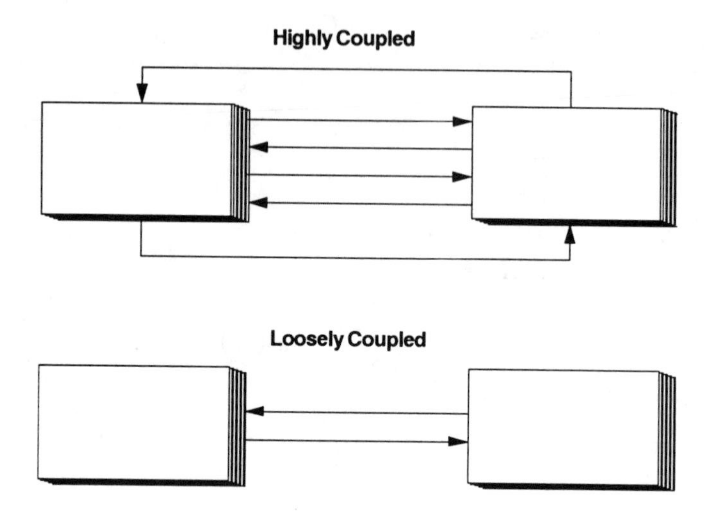


**FIGURE 13-2.** Module "coupling."

dence among the modules that make up programs. A major characteristic of functional independence is that interfaces are simplified. The minimization of interfaces is a highly desirable quality when a large team is involved in development. Functional independence can also result in code that is easier to maintain and debug. When modules are highly coupled, the probability of error propagation from one module to another is increased. Thus, a basic goal in detailed software design is to strive for modular independence or loose coupling.

A closely related concept is that of cohesion. Figure 13-3 depicts the basic concepts of low and high cohesion. A highly cohesive module performs a single task. A direct result of this high cohesion is that the procedure and data contained in a given module is inaccessible to other modules. This aspect, called information hiding, can greatly assist when software modifications are made. High cohesion is also of direct benefit in simplifying software testing. Good detailed software design strives for functional strength (i.e., high cohesion).

A related consideration at the module level is that of *complexity*. It is generally best to limit the function of a module to the simplest of tasks. For example, dividing what appears to be a simple functional task that contains two loops into two independent parts can greatly reduce combinatorial complexity during testing.

It should be stressed that none of these techniques inherently leads to program *correctness*. Correctness has to do with whether or not a program,

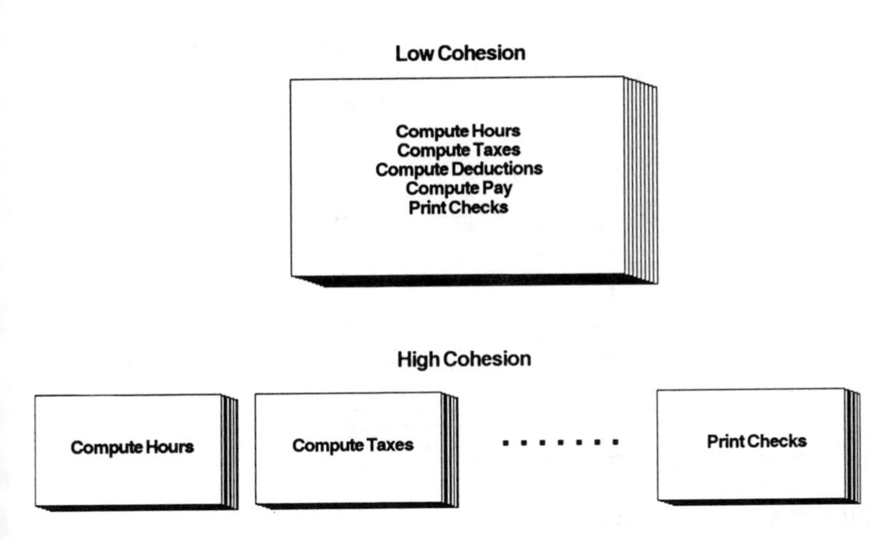


**FIGURE 13-3.** Module "cohesion."

consisting of modules, meets its specification. At a higher level, correctness has to do with whether or not the specification meets the actual user need.

## Data Structure Concepts

The complexity of programs that routinely access data bases can be profoundly affected by the choice of data structures. Data structures specify the logical arrangement of data elements. The choice of structure is important because it inherently sets the flexibility with which data elements can be accessed. A familiar case in point is the banking data base system through which a teller can access your account information solely through your account number, as opposed to information you are much more familiar with, such as your name. Data structures invariably affect the program design from its top-level structure to its modules to its lowest-level procedures. Tomes have been devoted to this important subject [1], [4], [7].

Pressman outlines five fundamental data structures [10]. These are shown in Figure 13-4. The size and formats of these classical structures are determined by the programming language.

The simplest of these, the Scaler element, consists of a single addressable data element that may consist of one or more bits, or a character string.

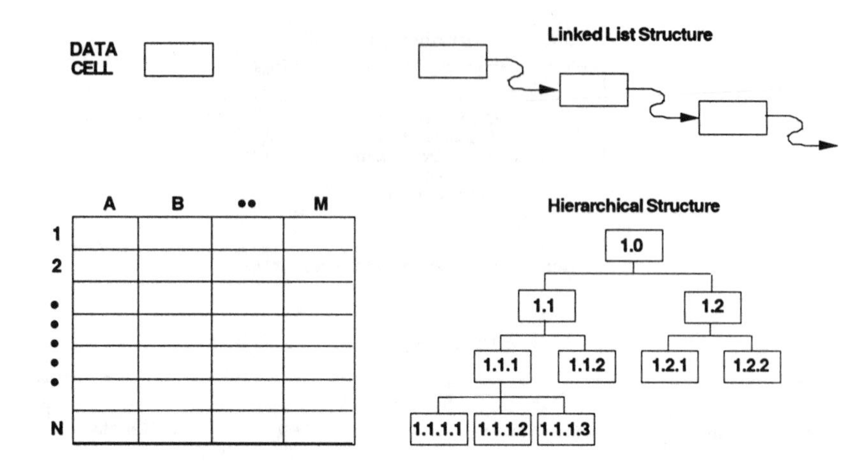


**FIGURE 13-4.** Fundamental data structures.

Vector elements consist of an adjacent set of scaler elements, or a list. They can be visualized as one-dimensional arrays, (A1:AN). Items in a scaler list are commonly sequentially accessed through indexed indirect addressing mechanisms, such as DO LOOPS or indexed BEGIN-END statements. Vector elements can be expanded to multiple dimensions to form arrays (A1:MN). Vector elements and arrays classically occupy contiguous space in memory.

Linked lists provide the capability to represent vector elements in non-contiguous memory locations. Using lists can result in more efficient memory use because memory is not allocated until it is needed, and then only as much as is required. Each scaler element contains pointers that may point forwards and/or backwards to the address of the next scaler element in the list. Stacks can be implemented using either vector elements or linked lists. Stacks are commonly built on a first-in-first-out (FIFO) basis, such as occurs in a queue, or a last-in-first-out (LIFO) basis.

Hierarchical data structures are composed of a number of linked lists. As the name implies, they are generally organized in a top-down fashion, such as class, kind, type, name, and attributes. For example, a resource class such as "transportation" may list kinds as "vehicles," "trains," and "aircraft." Types of vehicles may be "automobiles" and "trucks." Specific names and attributes then follow, such as the truck named "truck number 1" has the attributes "capacity = 2.5 tons," "current location = Freeport," and "range = 350 miles."

Hierarchical structures provide powerful mechanisms for grouping data entries and for data association. Grouping allows the collection of all kinds, types, and so on, such as the group of all TV sets or all automobiles. Association allows the formation of sets from different groups, such as all TV sets owned by high school graduates with incomes over $25,000, or all suspects over six feet tall who own blue four-door cars.

The classic simplicity of these structures should not lead to the impression that data base design is a simple matter. There is a strong tendency for data base structures to degrade through maintenance, particularly in the implementation of unforeseen user needs. The problem is greatly alleviated by knowing all the data that a particular application may require in advance. This, of course, is not always easy. It is also prudent to minimize constraints on data type generalization whenever possible. Data base maintenance has long been the bane of designers.

B. Walraet provides a good example in his discussion of a software system for a trucking company that defined a trip as an excursion taken by a single truck [12]. If a group of trucks was dispatched on the same itinerary on the same date, the event had to be treated as many trips. In short, the software requirements assumed a $1{:}n$ relationship between truck and trip, whereas the user really needed an $m{:}n$ relationship. This, of course, was discovered after the software was in use. The point is that the software designer would have done well to accommodate the relaxed constraint at the outset.

The example is simple, but instructive. Confronted with this newly discovered requirement, the programmer (who is constantly driven by motivations of performance) has the option to add a special routine to more efficiently accommodate the multiple use of the $1{:}n$ structure or to redesign the data base to accommodate an $m{:}n$ structure. What often happens is that performance wins and special tricks are employed. Soon, still new requirements emerge that, in turn, result in the methodic establishment of data redundancy and/or structural redundancy.

## DESIGN METHODOLOGIES

While the field of software design has significantly advanced since the 1960s, it remains a young and challenging enterprise. R. S. Pressman concurs in stating that software design methodology lacks the depth, adaptability, and quantitative characteristics that we typically associate with more classical disciplines, such as mechanical and electronic design [10]. Software design methodology, as opposed to simply programming and coding, is little more than two decades old. Still, methods for software design do exist and improved techniques are constantly emerging.

A central goal in the evolution of techniques for developing software requirements and design has been to devise an orderly process that exhibits:

1. A hierarchical structure detailing clear successive levels of design, thus enhancing communications, completeness, and the review process;
2. A logical division of functions and subfunctions (i.e., modular construction);
3. A capacity for ease in iteration; and
4. A capacity for the distinct definition of data, procedures, and terminology.

## Data Flow Diagrams

One widely used software design process intended to meet these goals is that of *data flow–oriented design.* The approach transforms information flow, represented in the form of *data flow diagrams* (DFDs), into a program structure.

DFDs consist of a set of top-down representations of data flow within a system. The representations are of a complete and successively detailed nature. The principles are similar to those of functional decomposition of a system, discussed in Chapter 5.

For example, consider a software system designed to provide ticketing reservations services. Figure 13-5 presents a simple representation of data flow at the top level. Entities outside of the system are shown in boxes. This is called the *context* level. Functions inside the system are shown in circles, and data flow is indicated by annotated arrows. Note that the flow lines do not cross each other. The context level DFD is deliberately simple. It's major purpose is to identify external users of the system and their informational interactions with the system.

Figure 13-6 provides an example of the level 0 DFD that presents the next cut at data flow representation within the system. Notably absent are the external entities that are generally shown at the context level only. The five circles represent the next level of functional decomposition. Each circle is numbered, serving as a designator for further DFD decompositions. Note that each entry within a circle begins with a verb.

The names given to each of the data flows portray the first attempt at constructing the *data dictionary.* The data dictionary is a highly structured,

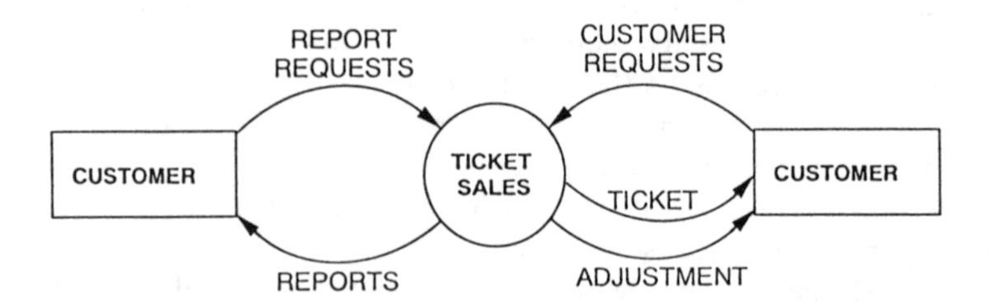


**FIGURE 13-5.** Ticket sales context flow diagram.

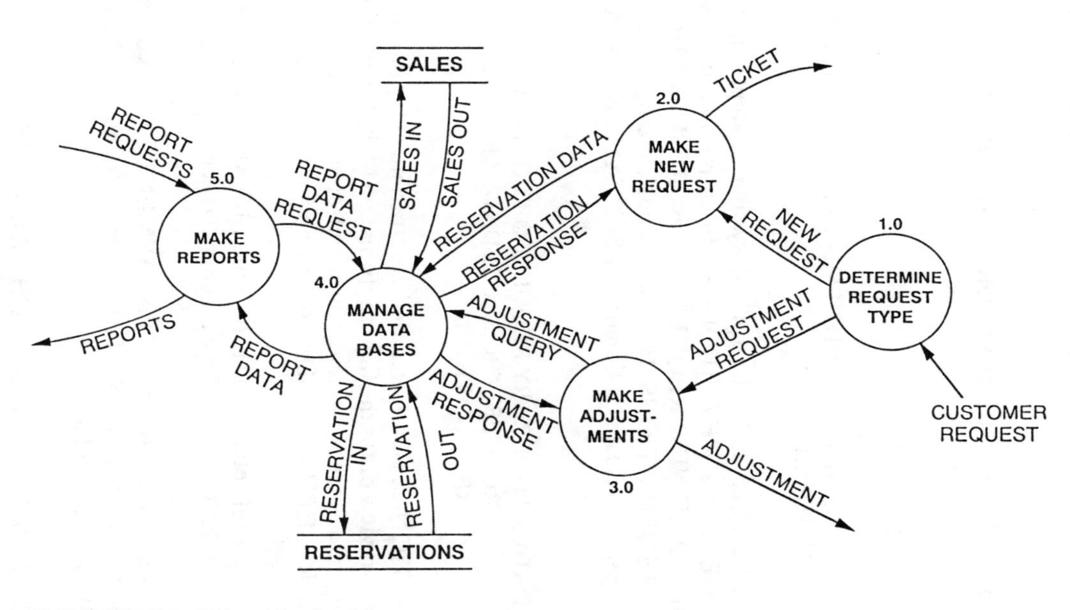


**FIGURE 13-6.** Ticket sales: level 0.

nonredundant, modular, and plain English description of each flow, files, data to be used within processes, and the elements of each. Figure 13-6 shows that there are two kinds of "customer requests"—a "new request" or an "adjustment request" to an existing reservation. In the data dictionary, this is denoted as:

CUSTOMER-REQUEST = NEW-REQUEST OR ADJUSTMENT REQUEST

and a "new request" is defined as:

NEW-REQUEST = PAYMENT-TYPE AND RESERVATION-DATA AND CUSTOMER-DATA

The DFD should be clear and self-explanatory. For example, if the CUSTOMER-REQUEST is a NEW-REQUEST for a reservation, data flow is directed to the MAKE-NEW-REQUEST process. RESERVATION-DATA flows to the RESERVATIONS data base through the MANAGE-DATA-BASES process. In the data dictionary, RESERVATION-DATA is defined as:

RESERVATION-DATA = ITINERARY-REQUEST OR CONFIRM-REQUEST OR (SALES-DATA AND PUT-RESERVATION)

That is, RESERVATION-DATA flow is used to extract itinerary information for the customer's review or for the final confirmation of the reservation or the actual making of a reservation and placement of sales data. These component flows of RESERVATION-DATA will appear in the flows of more detailed lower-level DFDs.

An example of lower-level detail is given in Figure 13-7, where the process number 2, MAKE-NEW-REQUEST, of the diagram 0 is decomposed to the next level of detail.

When the customer selects a reservation, the PAYMENT-TYPE is determined by the DETERMINE-PAYMENT-TYPE process. In the data dictionary, PAYMENT-TYPE is defined as:

PAYMENT-TYPE = (CREDIT-DATA OR CASH OR CHECK OR COUPON-DATA) AND CUSTOMER-DATA

After appropriate credit or flight coupon verification, the reservation is confirmed in the data base through the PAYMENT-VALIDATION flow and the RESERVATION-DATA flow. TICKET-DATA is then issued to the

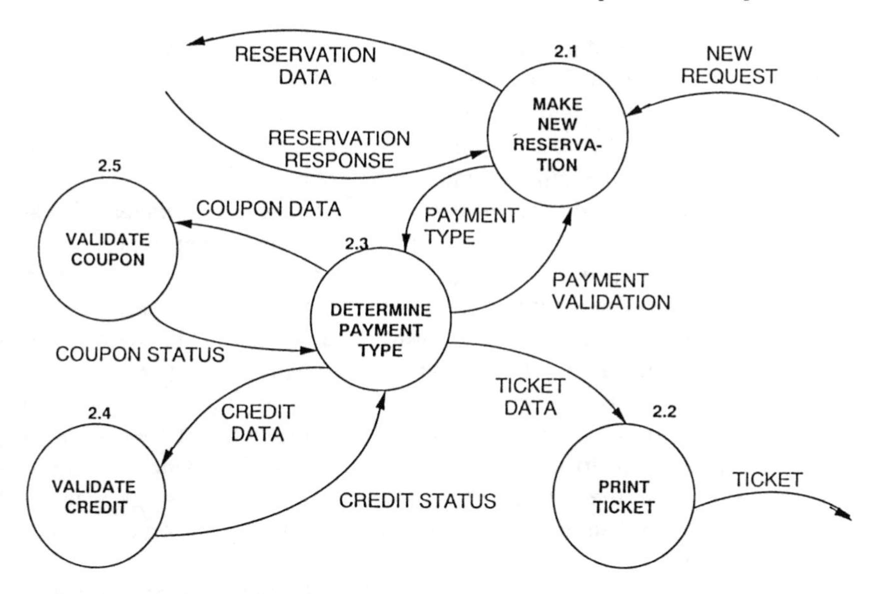


**FIGURE 13-7.** Make new requests: diagram 2.0.

PRINT-TICKET module. Note that the flow inputs and outputs shown in the level 0 diagram for the MAKE-NEW-REQUEST process number 2.0 (Figure 13-6) are the identical inputs and outputs shown in Figure 13-7 for the diagram number 2.0.

Processes depicted in circles are further decomposed in this manner into supporting DFDs, showing greater and greater design detail. It is recommended that any particular DFD employ no more than seven (at most nine) distinct circles, to maintain clarity. The DFD construct thus provides a means for hierarchical representation as well as logical functional decomposition. It also affords ease in iteration, of which a good deal is often required, and a methodology for precise definition of terminology— again, through considerable iteration.

DFDs are also routinely used to precisely define user requirements. The diagrams enhance communication between users and developers because omissions and errors are more obvious. In this capacity, DFDs provide an excellent tool for the review, iteration, and finalization of requirements with the software product end users.

## The Structure Chart

Structured design involves the systematic conversion of DFDs into structured charts. The program structure chart presents a comprehensive view of program organization. An example for the reservation system is shown in Figure 13-8. Each rectangle in the chart represents a program module. Each module has a name that clearly delineates the transaction that the module is to perform. Data flow between the modules is indicated by arrows with open circles. Arrows with filled circles denote control flows.

Structure charts can be *transform centered* or *transaction centered.* Transform-centered constructs are characterized by three general properties. These consist of an incoming data flow, a central transformation, and an outgoing data flow. The incoming data flow is processed (i.e., prepared for the transform function) and passed upward to a portion of the system that performs the basic transformation. Results are then processed (i.e., prepared for output) and passed downward to output interfaces. The three properties are isolated by examination of the DFDs, resulting in a boundary being drawn around the central transform. A structured chart is then built where the input data flow enters at the lower left, proceeds upward to a single central transform at the top center, and then descends on the lower right with the output data flow.

A large class of processes can be represented by the transform concept. These include data processing, simulations, engineering calculations, numerical analysis, process control, and so on. In fact, virtually all systems can be represented in this manner. In some cases, however, the concept of a transaction is clearly applicable, which gives rise to a different structured chart.

Transaction flow is useful when the data flow is characterized by a single incoming data item, which then gives rise to data flow along a number of different paths. The transaction flow representation is typified by data base inquiry systems, communications store and forward nodes, reservation systems, and banking transaction systems, to name a few. In these cases, the structured design charts are arranged around a transaction center that then fans out to descriptions of action paths, each of which serves a specific transaction type. There is no restriction on including both transform and transaction representations in a single-system design. The representation choice is largely up to the analyst. A basic consideration is one of readability and clarity.

Simpler programs are often designed using program design languages (PDLs) at the outset. PDLs provide structured English notation in pseudocode form. The PDL representation is similarly functionally decomposed into lower-level PDL specifications.

More complex programs, however, should adhere to the use of DFDs to develop requirements and preliminary design and structured charts for design. In this approach, PDLs are then typically used to transform structured charts into the next level of detailed design. In any case, it is noted that all of these techniques are basically implementation independent. An example of

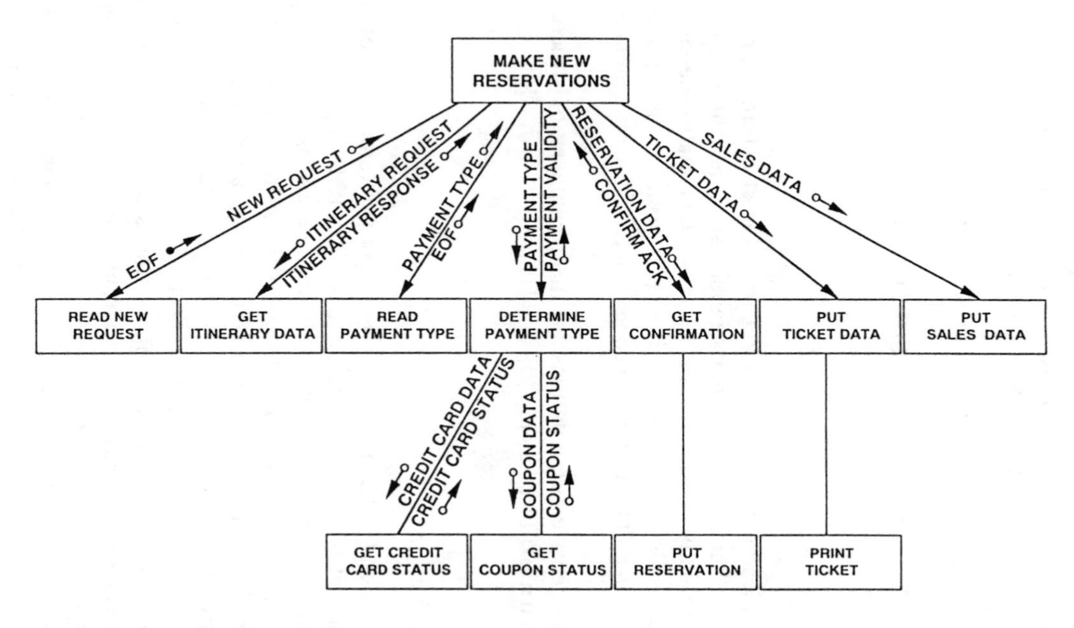


**FIGURE 13-8.** Program structure chart.

a PDL description of the DETERMINE-REQUEST-TYPE process in Figure 13-6 is as follows:

```
FOR EACH DATA STRUCTURE CUSTOMER-REQUEST
* Test Customer Request for New Request or Adjustment Request
      IF CUSTOMER-REQUEST = New Request
            THEN
                    LET CUSTOMER-REQUEST = NEW-REQUEST
            ELSE
                    LET CUSTOMER-REQUEST = ADJUSTMENT-
                      REQUEST
      END IF
END FOR
```

The PDL provides an orderly transition from the structure chart to a coding environment that is amenable to traceability and practical review. The pseudo-code is eventually translated to the actual implementation dependent code that has been selected. When possible, it is useful to use the same language structures in a PDL as those to be used in the final implementation language.

Further comprehensive presentations on the techniques of structured analysis and design are offered by [8], [9], and [11].

## Object-oriented Design

Object-oriented design (OOD) represents a distinctly different view of program design than that of structured techniques. The OOD approach views the software solution space as consisting of objects, operations, and messages.

An object is an easily visualized familiar element or function, such as an industrial process, files, a navigation computer, displays, a message switch, a carburetor, or any other device or process. An operation is a control or procedural entity that is performed by an object. The operation uses specific methods to accomplish the object mission. Methods within an object are invoked by the arrival of messages that move from object to object. Messages are requests for services from objects or system outputs from objects.

For example, if an object were an inertial navigation computer, its operation would be to continuously calculate latitude and longitude on the basis of an input message consisting of measurements from an inertial platform sensor (another object) and issuing outputs to a position-indicating object.

The architectural difference between object-oriented and conventional programming is shown schematically in Figure 13-9. In general, a message arrives at an object and invokes a method. A method may call on another

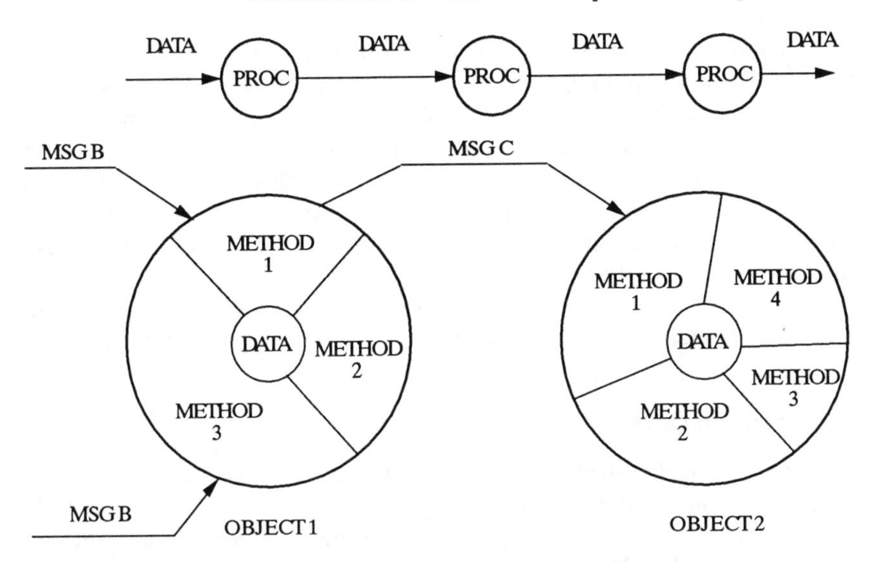


**FIGURE 13-9.** Object-oriented vs. conventional programming.

method within an object. Required data structures needed by a process or method are also contained wholly within the object's domain.

A major difference, then, is that in OOD the object contains its own private data structures and processes. The concept inherently provides a high degree of modularity. Details of implementation within an object are also hidden from other objects and from the outside world; thus, information hiding is built into the concept as well. As a result, as long as the method interfaces (arguments) are kept the same, the underlying implementation can change due to platform changes or changes in object data structures.

This same characteristic has led a number of analysts to suggest that OOD may offer a means to allow the software industry to move closer to a "manufacturing" business by eventually offering a set of premanufactured "parts," much as the hardware industry does today [6].

The OOD design process begins by partitioning the problem, in much the same manner as with other constructs. In this case, the analyst partitions the problem into objects, as opposed to carrying out the more traditional functional decomposition. After the objects and their attributes are identified, the operations they must carry out are defined. Interfaces are then established, detailing the relationships between objects and operations. As with other constructs, using structured English, PDLs, and/or pseudo-codes are useful in bringing structure and consistency to descriptions of operations and methods.

A major advantage of OOD is that problem isolation can be facilitated because the software embodiment is similar in structure to the human view of the real world structure itself. With OOD, a software bug within a process or method typically shows up as an incorrect message being sent to another object. This means that observing input and output messages for an object can often quickly isolate a faulty object. This feature is particularly useful when objects are defined along functional lines similar to those with which we view the real world process.

A clear exposition of the differences between OOD and functional programming, which is also rich in references, is offered by W. Booch [3].

## The Program Development Folder

The concept of the *program development folder* (PDF) is not new and, in practice, is employed with varying degrees of rigor. The term may even be "old hat" to some. The fact is, however, that the PDF or its equivalent is an extremely useful tool for the systems engineer in maintaining an understanding of progress throughout the software development process. When used properly, it is also an asset to programmers by providing a media to measure the completeness of each step and by focusing on the next tasks to be carried out.

The PDF basically provides a formal mechanism for the orderly assemblage and documentation of requirements, preliminary design, detailed design, and testing of software programs. Through each of these phases, the PDF should offer a clear definition of tasks to be accomplished, the sequence of those tasks, and a clear record of progress toward their accomplishment.

The exact content of the PDF, its use for review purposes and its advantages to all, should be an early subject of discussion by the SDT. As systems engineer, you will need to strike a balance between the need for consistent recording of development status and excessive imposition of paperwork on the programmers.

A suggested PDF document content is given in Figure 13-10. The *cover sheet,* shown in Figure 13-11, identifies the programmer, the software covered by the PDF, and the PDF custodian. The custodian of a given PDF is normally the programmer or the programmer's immediate supervisor. The cover sheet also lists the contents by section and provides space to enter a due date and date completed for each section.

The requirements section begins with a top-level English description of the requirement(s) to be met by the program and should reference the specific requirement(s) documented at a higher level. The English description is then followed by a series of top-down data flow diagrams that meet requirements at the next levels of detail.

| SECTION | CONTENT |
| --- | --- |
| 1.0 | Requirements |
| 2.0 | Design Description |
| 3.0 | Program Code |
| 4.0 | Program Test Plans |
| 5.0 | Test Case Results |
| 6.0 | Deviation Log |
| 7.0 | Notes/Appendices |

**FIGURE 13-10.** Program development folder contents.

Program Name ____________________ Programmer's Name _______________

Custodian ___________________________

Component/Modules included ___________________________________________

_____________________________________________________________________

| SECTION NUMBER | DESCRIPTION | DUE DATE | DATE COMPLETED | ORIGINATOR |
| --- | --- | --- | --- | --- |
| 1.0 | Requirements | | | |
| 2.0 | Design Description | | | |
| 3.0 | Program Code | | | |
| 4.0 | Program Test Plan | | | |
| 5.0 | Test Case Results | | | |
| 6.0 | Deviation Log | NA | NA | |
| 7.0 | Notes | NA | NA | |

**FIGURE 13-11.** Program development folder cover sheet.

The program design is documented in the second section, using project-adopted structured diagrams. The description of the design proceeds with translation of the structured diagrams into pseudo-code, using a PDL. The actual code, which is an extension of the design, is included in the next section in the form of the current source code listing.

The due date for this entry corresponds to the first error-free compilation of the program.

The PDF is a living document throughout this process. When concluded, it should be suitable for direct inclusion into the software design specification, with minimum revision. Design walk-throughs are normally held at the completion of the requirement DFD descriptions, structured diagrams, and pseudo-code production stages.

The program test plans contain a description of the testing approach for the program. This section identifies required software drivers and/or tools to be used, test inputs, and expected outputs. Tests are planned to meet the requirements specified in section 1.0 of the PDF, which are traceable to higher-level requirements. In addition to functional testing, appropriate tests should be devised for error handling, range values, and logical path analysis.

Section 5.0 contains a description of successful test results. Section 6.0 contains a record of any revisions to requirements, design, test plans, drivers, and tools required to achieve successful implementation of the program. Section 7.0 includes any additional information deemed necessary to clarify all aspects of the program implementation for an independent reviewer.

The degree to which the systems engineer imposes rigor in the use and content of the PDF is directly related to the degree of his or her confidence that implementation of given programs will be successful. The lower the level of confidence (for any reason), the greater the need for visibility. Typical reasons for an increase in the need for visibility are the use of large programming staffs, programming teams that are organizationally or physically separated, and any issues related to technical feasibility. In these and similar instances, the use of a PDF (or an equivalent disciplined approach) can provide an important management tool for both the software systems engineer and the systems engineer. The PDF concept can also be used at lower levels of program design, including computer program components (CPCs) and even modules as required.

When visibility is required, the major advantages of the PDF are:

- There is a clear and consistent understanding of which individuals are accountable for and specifically for what.
- The PDF provides management visibility at any point in time, independent of reviews and walk-throughs.
- Design and test documentation is developed as it takes place.

- The PDF provides a mechanism for traceability.
- Communication among programmers (interface understanding) is enhanced.
- The PDF provides a mechanism for recording change as it occurs.

## WALK-THROUGHS AND PEER REVIEW

The walk-through is an in-depth technical review of the progress of software design and/or implementation. Walk-throughs are basically mini-reviews conducted on specific parts of the overall software system. Their frequency and timing is solely up to the software systems engineer and the systems engineer and is determined by the need for visibility. Schedules for walk-throughs are coordinated through the SDT under the systems engineer's leadership. Additional walk-throughs may be planned at any time.

The walk-through team should consist of four to six people as a rule. These include the systems engineer, the software systems engineer, and selected peers with direct interfaces with or knowledge of the program under review. Presentation material should be standardized and distributed to attendees, with sufficient time for review prior to the walk-through—at least one week.

It should be emphasized that a walk-through is not a performance evaluation. It is actually an opportunity for the reviewee to greatly enhance the probability of his or her success in software development.

Walk-throughs can be held to review requirements, design, code, test, or at any other point considered necessary. Walk-throughs are chaired by the software systems engineer or an appointee. Action items are assigned to responsible individuals, with specific, realistic due dates. The following paragraphs cover the major considerations to be addressed at requirements, design, and code walk-throughs, respectively.

### Software Requirements Walk-through Topics

- A statement of program functional requirements in the context of total system partitioning.
- An English language summary of derived requirements.
- A listing of derived requirements, with traceability of each to the system functional requirements by paragraph number.
- A data flow diagram representation of requirements, including a clear description of interface requirements.
- A current version of the data dictionary in support of the DFDs.
- Are the requirements achievable both technically and programmatically?

- Are the requirements testable, and what method will be used for their testing?
- Is the PDF up to date?

### Software Design Walk-through Topics

- A complete design description, using structured programming and project approved structured tools.
- The ability of the algorithms to meet required functions and computational error envelopes.
- Is the design traceable to the requirements?
- A description of all data structures, data bases, and the data dictionary.
- A description of all error handling routines.
- A clear description of interface design.
- A report on margin status as appropriate.
- Is the PDF up to date?

### Software Code Walk-through Topics

- Does the code comply with the design?
- Are comments sufficient for the unfamiliar reader?
- Is the language efficiently and properly used?
- Are data constants, typings, and declarations correct?
- Is the PDF updated and complete to date?

Aside from the mechanism of the walk-through, an extremely effective method for checking code is simple peer review. This is accomplished by asking programmers on the project to set time aside to read each others DFDs, structure diagrams, and code in isolation. It is often difficult when involved at levels of considerable detail to see simple errors or to provide adequate comments when dealing with material that is so familiar and obvious to a programmer. Peer review provides an opportunity for a truly fresh look at the work. Finding errors through peer review is especially valuable in that a relatively small investment of time at this level can save significant amounts of time if the errors need to be corrected at the integration and system test levels of effort.

#### References

1. Aho, A. V., J. Jopcroft, and J. Ulmann. 1983. *Data Structures and Algorithms*. Reading, Ma.: Addison-Wesley.
2. Bergland, G. D. 1990. A Guided Tour of Program Design Methodologies, *Milestones in Software Evolution*, IEEE Computer Society Press.

3. Booch, G. 1986. Object-oriented development. *IEEE Transactions on Software Engineering,* Vol. SE-12, No. 2, February.
4. Date, C. J. 1986. *An Introduction to Data Base Systems.* Addison-Wesley.
5. Dijkstra, E. W. 1968. Communications of the Association for Computing Machinery, March.
6. Haavind, R. 1992. Software's new object lesson. *Technology Review,* February/March.
7. Kruse, R. L. 1984. *Data Structures and Program Design.* Englewood Cliffs, N.J.: Prentice-Hall.
8. Page-Jones, Meiler. 1988. The Practical Guide to Structured Systems Design. Yourdon Press, Prentice-Hall.
9. Peters, L. 1987. *Advanced Structured Analysis and Design.* L. Englewood Cliffs, N.J.: Prentice-Hall.
10. Pressman, R. S. 1987. *Software Engineering: A Practitioners Approach.* New York: McGraw-Hill, p. 213.
11. Teague, L. C., Jr. 1985. *Structured Analysis Methods for Computer Information Systems.* C. W. Pidgeon, Science Research Associates.
12. Walraet, B. 1988. *Programming, the Impossible Challenge.* Amsterdam: North-Holland.

# 14

# Testing

Early testing is important because detecting and correcting errors during system integration and test can cost up to 8 times the cost of detecting and correcting errors at the preliminary design stage, and up to 100 times the cost of detecting and correcting errors in operations. From the systems engineer's standpoint, testing in support of system development is structured around four major conceptual activities. These are unit testing, system integration and test (SI&T), system testing, and acceptance testing.

Because early testing can be of such importance, the systems engineer should routinely involve test engineering representation on the SDT throughout the entire systems engineering process. The ability to test a system can have a major impact on its design.

Establishing the system test requirements from which detailed SI&T, system, and acceptance tests should be derived is also a responsibility of systems engineering. Testing approaches and detailed processes are delineated as *test plans and procedures* (TP&P).

The TP&P are formal documents derived from system test requirements. The plan part of the TP&P document specifies how each specific test is to be implemented. The plan specifies the test verification approach, equipment to be used, types of personnel required, an overall schedule, anticipated results, the format for test report documentation, and a clear traceability matrix to each functional requirement as a minimum.

There are four basic verification techniques:

1. Inspection—Consists basically of visual examinations of engineering drawings, software designs and/or listings, configurations, and so on.
2. Analysis—Use of simulation and/or analytic modeling and other analyses to verify performance limits and accuracies. Analysis is used when direct testing is deemed to be too complex, time consuming, or costly.

3. Demonstration—A verification method less stringent than detailed test used to illustrate top-level functional go/no-go capabilities.
4. Test—Employed to verify requirements through the use of equipment, precise measurements, and detailed procedures.

The procedure part of the TP&P specifies in detail the exact sequencing of events, specific daily schedules, required instrumentation, configurations, support equipment, personnel assignments, complete traceability mechanisms for validation of requirements, and expected detailed results, with appropriate reporting forms.

For example, the detailed procedures for radio propagation contour measurements may include driving an instrumented vehicle from point A to point B on a specific day and measuring signal strength at input to the first IF stage at specified time intervals. A software test example might be to insert an exact test word in a specific register, execute a particular code path to a breakpoint, and examine a predicted bit pattern in a second register. Detailed test report forms include anomaly reporting and accommodation for retesting and reporting.

TP&P documents are routinely developed for SI&T, system testing, and acceptance testing. They may also be developed for unit testing if testing called for at the unit folder level does not adequately prepare for coordinated unit entry into the system integration and test phase.

## UNIT TESTING

Unit, or subsystem, testing is a work item under the development of mission product items and is the responsibility of subsystem COGEs. In this scheme, fully tested and functioning mission product items are submitted for integration testing. While unit testing is primarily within the province of the COGE, there are still many system-level issues to be considered. The ability to test a unit has a strong impact on its design, as does the ability to test it at the SI&T and system levels. These testing strategies should be coordinated so as to make use of similar validation and verification methodologies as often as possible. Developing such strategies is a system issue.

The systems engineer should also be aware of the many testing strategies that can productively be carried out below the subsystem and program levels. These include complete or probabilistic parts testing upon receipt or fabrication, subassembly and component testing, and thorough testing of critical software algorithms, procedures, modules, and CPCs.

## SYSTEM INTEGRATION AND TEST

System integration and test (SI&T) consists of the methodic interconnection and testing of subsystems in increasing numbers and in an increasingly complex manner until the entire system is interconnected and its interfaces completely checked out. SI&T marks the first time in the development cycle where the systems engineer and everyone else is able to gain hard evidence of the probability of success. While other devices, such as prototyping, simulation, structured code, peer reviews, and code walk-throughs, may have been employed prior to this point to enhance confidence in the integrity of the design and fidelity of the implementation, this is really the first time in the development process that the actual system initially comes together—and works or doesn't work.

It is a time that can be fraught with disaster and stymied with much finger pointing, particularly when multiple vendors or organizations are taking part. It is also a time when the professionalism and cooperation of all parties involved can be strenuously tested. One rule of thumb in developing schedules for SI&T is to take the best estimate you and your associates can make, and, when you are all comfortable with your guesses, multiply the time line by two (some say three!).

The major reason that SI&T can be such a lurking problem is that, when a deficiency is found and corrected, the correction needs to be retested. A basic question that is always confronted at this point is whether to perform this retest in isolation (i.e., test for the exact fix) or to repeat integration testing to that point to insure that the fix has not introduced other subtle discrepancies. The problem is most severe in software testing, where such subtleties are often not apparent.

The repetition of integration testing is called *regression testing*. In its simplest form, regression testing requires the complete resetting of the SI&T clock every time it is performed. While partial regression testing is less demanding, it is still extremely difficult to anticipate the magnitude of retesting as well as the exact methods to be employed for retesting at the project planning stage.

There are very meaningful steps that can be taken in system and subsystem design to guard against the spiraling chaos that SI&T can easily degrade to. When something goes wrong, the whole name of the game is fault isolation and rapid retest. In hardware, it is difficult to overestimate the importance of strategically located, functionally oriented test points. It is very important to instill this awareness in each designer's thought process, even to the point of conducting special internal reviews on the subject. If you can afford it, it is better to have too many test points than to wind up saying, "if I only had a test point—there!"

An early, extensive, and well-planned component, subassembly, and unit test prior to SI&T can also greatly reduce malfunctions at this time. Identifying faulty components at the component level is considerably cheaper per fault than identifying at the SI&T level.

There is also a distinct trade-off between the size of subassemblies to be tested and the complexity (hence costs) of the tests themselves. The decision of what goes into a subassembly directly impacts the modularity of the design. In some cases, testing alone can dictate subsystem partitioning itself. There is no doubt that, in dealing with highly complex systems, experienced test personnel are a significant asset to the SDT. The payoff is often more than evident in SI&T.

In software design, functional modularity and high cohesiveness are key features in augmenting ease in fault isolation. A major virtue of *object-oriented programming* lies in the relative ease of associating functional malfunctions with specific functional modules. Still, one of the most effective methods of early fault detection in software—a method seldom practiced—is simple peer review of source code prior to compilation. Code and design walk-throughs, especially in areas where your confidence may be less than total, are also a must.

From the day you begin a project as a systems engineer, you must constantly bear one thing in mind. One day you will reach SI&T. Extensive, unplanned regression testing is without question the single biggest threat to your schedule. It has been termed the "black hole" of software development. When testing begins to drag out beyond the "planned" time, there is increasing pressure to get the job done. Tough decisions have to be made to minimize further delays. Complete regression testing can easily give way to partial regression testing and then to simple verification of isolated fixes. The worst comes when your TP&Ps are degraded or even abandoned. Under these pressures, configuration management can become a nightmare. Not far behind walks your credibility.

The influence of testing upon system and subsystem design is a significant one. Give this matter its due attention by heightening the awareness of your entire team to these issues, and, should there be any doubts, make good use of experienced test personnel early in the design, without hesitation.

## SYSTEM TEST

There may seem to be a subtle distinction between SI&T and system-level testing. SI&T deals with the methodic sequential assembly and testing of system and subsystem interfaces. While system interfaces are, in fact, interfaces of specific subsystems to the external world, and hence may have been tested to some degree in SI&T, the emphasis of system testing is on system-level responses to system-level inputs—that is, the response across all subsystems that a particular test path may take.

System testing is carried out by the implementing organization prior to any customer involvement in testing. The purpose is to verify that all system-level requirements are met by the implementation of the system design. These include all end-to-end information flow requirements as well as environmental, performance, and system availability requirements.

As in all testing, the system test plans are derived from the system functional requirements document, and the test procedures are derived from the test plans. In both cases, traceability to all functional requirements must be shown as an integral part of the plans and procedures.

When the implementation organization is satisfied with system-level testing, it is basically ready to begin customer acceptance testing. It is common to conduct acceptance testing at the user's site. System testing, therefore, is often divided into two distinct iterations—the first prior to system breakdown and shipment to the vendors site(s) and the second after shipment and installation at the user's location. Ideally, system-level tests at the vendor's site prior to shipment should take place using the *identical* configuration that will be used at the user's site. This includes such details as the same placement of equipment, use of cables, environment, and so on. Pre-ship and post-ship reviews are often scheduled to insure the readiness and adequacy of these system-level tests.

Deviations from this course may arise if it is expedient to deliver parts of the system directly to the customer's location rather than to a central vendor site for system testing. In these cases, final stages of SI&T may actually be accomplished at the user site prior to system testing.

## ACCEPTANCE TESTING

When the developing organization is satisfied that the entire system operates in a manner that meets all functional requirements, it is then ready to convince the customer that this is indeed the case. The purpose of the acceptance test plan is to establish a well-defined, previously agreed to point at which the system can be transferred to the customer. Successful completion of acceptance testing and the subsequent signing off of that completion defines the point at which the implementation is officially completed—the same point at which operations and maintenance contracts usually begin.

It is very important to seek early agreement between the vendor and the customer on the philosophy of how the system will be accepted and on the details of actual acceptance testing. These TP&P for acceptance testing need to be known to the systems engineer so that adequate planning can take place in the design of system integration and system level testing prior to acceptance testing.

In some cases, acceptance testing is merely a repeat of the implementing organization's system-level tests, with the customer participating and/or observing. In all cases, acceptance testing should be a subset of the system-level testing previously carried out. Any other arrangement can easily lead to misinterpretations, which is why the systems engineer must negotiate the exact nature of acceptance testing in advance of all implementation test planning.

In simpler systems, the customer may execute an entire repeat of system-level testing, with the implementor's support. In complex systems, other realistic strategies may be required. For example, consider an actual case of a software system designed to carry out automated testing of inertial navigation computers at the board level upon the completion of manufacture. It was agreed that the acceptance test for the fault isolation software would consist of a number of software test runs. Prior to each run, a tester from the customer organization would remove a TTL package from a board known to be good, bend one of the pins upward, and replace the chip back in its socket so that the affected pin did not make contact. The software under test would then be run to see if the fault was successfully isolated to that particular chip. It was further agreed that carrying out this procedure for every pin on every chip on every board would require an excessive amount of time. The customer agreed that, if his personnel could carry out this procedure randomly 8 hours a day for a period of 2 weeks without encountering an error in the fault isolation software, this would constitute grounds for formal system acceptance. This statistical test was carried out successfully, and the system was accepted after the first 2 weeks of testing. The important point is that the approach was negotiated early, before the PDR, and this knowledge impacted the test design throughout implementation.

Measuring performance parameters, such as response times, error rates, propagation coverage, acceleration, and other related functional requirements, are also statistical measurements requiring negotiated, realizable methodologies. Simulation of external stimuli to the system are often required.

Measurement of system availability, another statistic, is often accomplished over an agreed-to period in an operational setting. A typical approach to availability testing is to reset the test period clock upon correction of each failure that falls short of the desired *mean time to failure* figure. This approach is favorable to the customer. Limits to the number of resets to be tolerated must be worked out in the interest of both parties.

Details of acceptance testing are often overlooked until late in the implementation process, perhaps because it seems so far away at the beginning of an intricate and lengthy project. Be aware of the importance of addressing this issue early to avoid unforeseen requirements and attending schedule alterations. It is an area of great interest to the systems engineer, as it defines the criteria for achieving the end point of all of his or her efforts.

Adequate testing of commercial systems designed for use by a large number of customers presents an additional problem. One approach to this circumstance is the use of alpha and beta testing concepts. In alpha testing, the product is taken to a specific representative customer, where testing takes place as a joint effort between the developer and the customer. Beta testing involves the distribution of the product to a number of prospective customers, who each test the product in their own operational settings.

A combination of the two strategies is often used, where the alpha test is designed to reveal the majority of customer recommendations and the beta test is then used to refine these findings over a larger sample base.

## LESSONS LEARNED

Experience ordains that the following factors directly related to testing are among the most predominant causes of unforeseen problems. The problems range from development schedule and cost impacts to increased operational costs to outright operational failure.

1. *Inadequate low-level testing*—Refers to the failure to conduct adequate testing of hardware parts, subassemblies and components, and software procedures, modules, and CPIs.
2. *Inadequate consideration of test impacts on design*—Lack of sufficient test points, with regard to strategic function and/or number. Usually occurs when too many parts are placed on a single board or platform. This is an assembly and subassembly partitioning issue.
3. *Conservative environmental tests*—Negligence in executing tests over the complete range of environmental extremes over a sufficient period of time, in accordance with functional requirements.
4. *Deviation from test plans and procedures*—This most typically happens when schedule and cost limitations are encountered due to unforeseen extensions in regression testing. Programmatic pressure results in the subordination of adequate retesting at system levels. Be aware that the staircase and staircase with feedback systems engineering paradigms are not only most susceptible to this condition, but it is an extremely common occurrence attending the use of these models.
5. *Inadequate early planning for test logistics support*—This oversight involves the discovery that additional test equipment or other testing re-

sources, such as software drivers and stubs, are required in addition to those originally planned for in the TP&Ps. Other logistics oversights include inadequate planning for test equipment failures and the logistics of pre- and post-shipment accommodation details.

6. *Insufficient aggressiveness in testing*—Primarily motivated by the overwhelming desire to pass, which commonly supplants the more intelligent mind-set that the finding of errors, in the long run, is a positive event.

# 15

# Systems Engineering Management Aids

Successful technical managers are successful because they have technical visibility. In the context of systems engineering management, visibility is that quality that enables its possessor to foresee problems early enough to devise and implement corrective actions at lower levels of scheduled activity, such that major project milestones are not impacted.

This chapter discusses a number of diverse tools, strategies, and procedures that taken individually may seem unrelated. From the mind-set of the systems engineer, however, these aids are strongly related in that their fundamental purpose is a common one. That common purpose is to employ organization, structure, sound processes, and sensitive human relations to provide a coordinated course of action, the sum of which contributes to a punctual identification of problems—that is, visibility.

A basic checklist of systems engineering management aids that provide such visibility is given in Table 15-1. Some of these issues are logically discussed elsewhere in this text. They are assembled here for completeness. Chapter references are provided where appropriate. Those not covered elsewhere are discussed in more detail here. While the list may be expandable, it is certainly fundamental. The important message in reviewing these aids is to understand that all of these techniques (and similar ones the reader may wish to include based on his or her own experience) can be viewed as commonly serving to support a chronic need of the systems engineer—the maintenance of vision.

## CHOOSE THE RIGHT PARADIGM

Four basic paradigms for the systems engineering process are presented in Chapter 2. These models consist of the staircase with feedback, early proto-

**TABLE 15-1 Systems Engineering Management Aids (Aids for the Maintenance of Visibility)**

| |
|---|
| Choose the right paradigm |
| Performance measurement |
| Reviews |
| A sound work breakdown structure |
| Margin identification |
| Configuration management |
| Adequate testing |
| Personal skills |
| Organization |
| The system design team |

type, spiral, and rapid development models, and/or modified combinations of these. Selecting a development strategy at the outset that does not fit the reality of the project you intend to undertake will greatly increase the probability of unanticipated major disruptions down the road.

As the element of uncertainty increases, the systems engineer should favor adopting the spiral and/or rapid development models. Be aware, however, that these models are still new to many organizations and that management standards for these models are still evolving. Sound, sensible, strong, and experienced technical management is a definite asset when using the newer paradigms.

Also, independent of the paradigm selected, successful system development requires continuous and intensive user interactions through permanent and responsible membership of the user community, or competent representation thereof, on the SDT.

Give careful thought to mapping your project into and onto the right paradigm.

## PERFORMANCE MEASUREMENT

The major purpose of performance measurement is to understand progress to date relative to an original plan, identify shortcomings, provide insights into the future, and aid in suggesting corrective action(s). Perhaps the simplest method for tracking achievement is the classic Gantt chart, where tasks are laid out on time lines. The familiar Gantt chart includes notation for tracking completeness, anticipated end dates, extensions for tasks, and due dates for products that result from tasks.

Gantt charts are widely used and are particularly useful for monitoring

| | ACTIVITY | PREDECESSOR ACTIVITY | TIME ESTIMATES (WEEKS) $T_O$ | $T_M$ | $T_P$ |
|---|---|---|---|---|---|
| B. | BUILD FACILITY | NONE | 20 | 24 | 30 |
| F. | SAFETY INSPECTION | B | 2 | 3 | 4 |
| C. | INSTALL EQUIPMENT | B | 8 | 16 | 20 |
| D. | RECRUIT WORKERS | NONE | 2 | 2 | 3 |
| E. | TRAIN WORKERS | NONE | 4 | 5 | 6 |
| A. | PERFORM PILOT | C, E, F | 4 | 5 | 9 |

**FIGURE 15-1.** Network scheduling—CPM/PERT.

weekly progress at design team meetings. The major advantage of the chart is that it is easy to read and serves as a convenient basis for discussion at lower schedule levels. The principal disadvantages of the Gantt chart are that it does not clearly show interdependencies and it does not consider cost.

A common method for clarifying task interdependencies uses the program evaluation and review technique referred to as PERT, [1]. PERT was originally developed in 1958 for the Polaris missile project under U.S. Navy sponsorship. The approach is summarized in Figures 15-1 through 15-4.

Results of the first step in the use of PERT are shown in Figure 15-1, where the predecessor activities for each required activity are listed. Also listed are estimates for the most optimistic times ($T_O$), or shortest times, to complete each task, along with the estimated mean times ($T_M$) and the most pessimistic, or longest times, ($T_P$) to complete each task.

The first activity in the sample schedule, activity B, is the construction of a facility. Executing this activity does not depend on any other activity. Activity F, that of safety inspection, cannot begin until activity B is concluded.

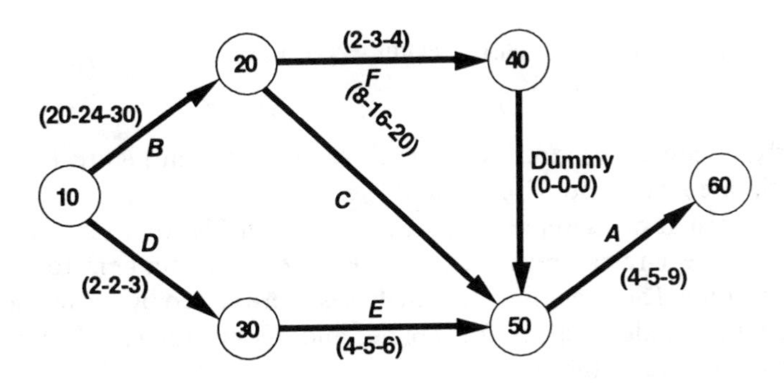


**FIGURE 15-2.** Network scheduling—CPM/PERT (cont.).

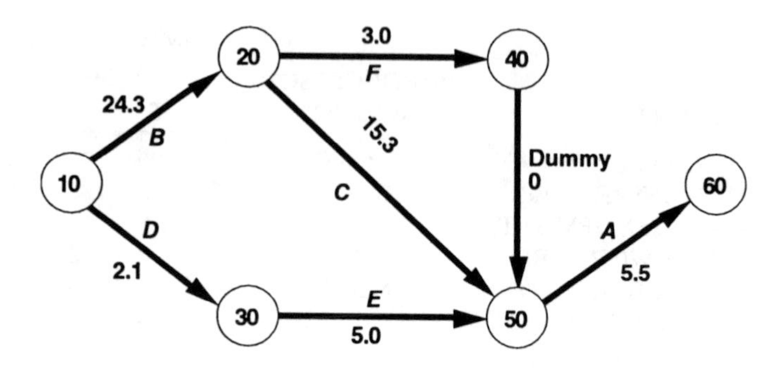


FIGURE 15-3. Network scheduling—CPM/PERT (cont.).

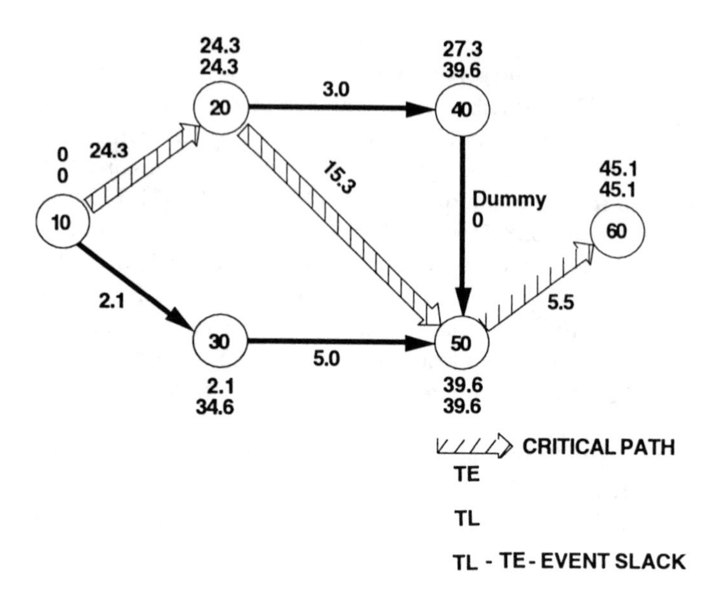


FIGURE 15-4. Network scheduling—CPM/PERT (cont.).

Similarly, dependencies for the other tasks are noted in Figure 15-2, along with estimated values for $T_O$, $T_M$, and $T_P$.

This information is represented graphically in Figure 15-2. The values entered in the circles are arbitrary node reference numbers to facilitate discussion only. The execution of activities is represented by arrows between these circular nodes. Values for $T_O$, $T_M$, and $T_P$ are indicated along each activity path in parentheses.

Each activity is directed to a unique node. Since activity A cannot begin until activities C, E, and F are all completed, it is convenient to note the completion of these three activities at the single node numbered 50. Dummy activities are used in PERT charts to effect this closure.

The next step is to develop an estimate, $T_e$, for the expected time to complete each activity. The PERT convention for determining this estimate is given by

$$T_e = \frac{[T_o + 4 * T_M + T_P]}{6} \tag{15-1}$$

which can be viewed as a mean value with weighting on the $T_M$ term.

Figure 15-3 displays the results of the calculations for $T_e$ for each activity path.

It is clear that the longest time path from node 10 to node 60 consists of passage through nodes 20 and 50. This path constitutes the critical path for the entire project. Figure 15-4 highlights this critical path and also notes derived values for the earliest expected times, *TE,* and the latest expected times, *TL,* that an activity can expend in reaching the node it is directed toward. For example, the earliest expected time that activity completion to node 40 can occur is just the sum of the expected times for activities B and F, or 27.3 = 24.3 + 3.0. Note, however, that activity F does not really need to be completed until the critical path activities of B and C are completed. This is because activity A, the last activity, cannot begin until activities F, C, and E are all completed. This means that activity F can take as long as activity C, without jeopardizing the schedule—an amount of time equal to 15.3 time units. Thus, the longest time, *TL,* that can expire in reaching node 40 is 39.6 = 24.3 + 15.3.

Figure 15-4 also shows similar calculations for the remainder of the network. The difference between a given value for *TL* and *TE* (*TL* − *TE*) is referred to as the *event slack*. Note that the event slack along the critical path is simply equal to the expected time along the critical path—that is, there is no extra time to spend. The PERT concept clearly points out those activities that require the most management attention and provides a valuable tool in identifying high-risk areas.

Neither the Gantt Chart nor the PERT Chart as discussed specifically provide for the tracking of costs. The easiest way to do this is to select those items for which cost visibility is deemed important and simply track actual costs incurred versus scheduled costs for each item and for totals. Tracked items may include costs for personnel, procurements, travel, services, overhead, and so on.

There are at least as many ways to do this as there are organizations. A wide variety of commercial software packages are also designed to support this level of costing. Spreadsheets also furnish a convenient method for building cost profiles and provide quick plotting capability for ease in observing trends.

A more sophisticated approach designed to clarify the relationship between expenditures and the actual progress of work at a given point in time is offered by the cost, schedule, and performance tracking concept—commonly referred to as *earned value.* In this scheme, variances in both cost and schedule are measured in terms of planned and actual expenditures. Three new terms are introduced for these purposes:

BCWS—The *budgeted cost of work scheduled* (BCWS) is the estimated cost to perform the required work based on the original project schedule.
BCWP—The *budgeted cost of work performed* (BCWP) is the original budgeted estimate of the actual work performed to date.
ACWP—The *actual cost of work performed* (ACWP) is the real cost of the actual work performed to date.

Each of these terms has its units in dollars and can be plotted as a function of time. Figure 15-5 gives an example of how these concepts interrelate. The dotted line in the figure shows the profile of the cost as originally planned. The solid line tracks the actual cost incurred for the actual work performed at the reporting date. The lower dashed line shows the original budget that was allocated for the actual work completed at the report date.

In this formulation of data, the difference between the BCWP and the BCWS can be viewed as a measure of schedule variance. In the example of Figure 15-5, the schedule variance suggests that the work is behind schedule because the budgeted cost of the work performed is less than the budgeted cost of the work scheduled to be completed at the reporting time. There is also a cost overrun because the actual cost of work performed is higher than the amount budgeted. These discrepancies in terms of schedule and cost variance are added to the original BCWS curve, to arrive at a new estimate of schedule completion and required costs. The approach assumes that the BCWP is a faithful measure of the actual work performed. When this is true, the concept can provide a single performance measurement framework in which to view the relationships between cost and work completion.

There should be a structured consistency between items included in the work breakdown structure and the items that appear on schedules, the items upon which performance measurement is carried out and the items for which costs are tracked. While each of these functions may be carried out at different levels, depending on the project and the particular issues that the systems engineer and/or management wishes to focus on, the selected items should be traceable to the basic definition of the work to be performed (i.e., the WBS). Interestingly enough, this is commonly not the case. Accomplishment this compatibility is greatly facilitated by the use of consistent terminology. For example, the names of items that appear on a schedule should be

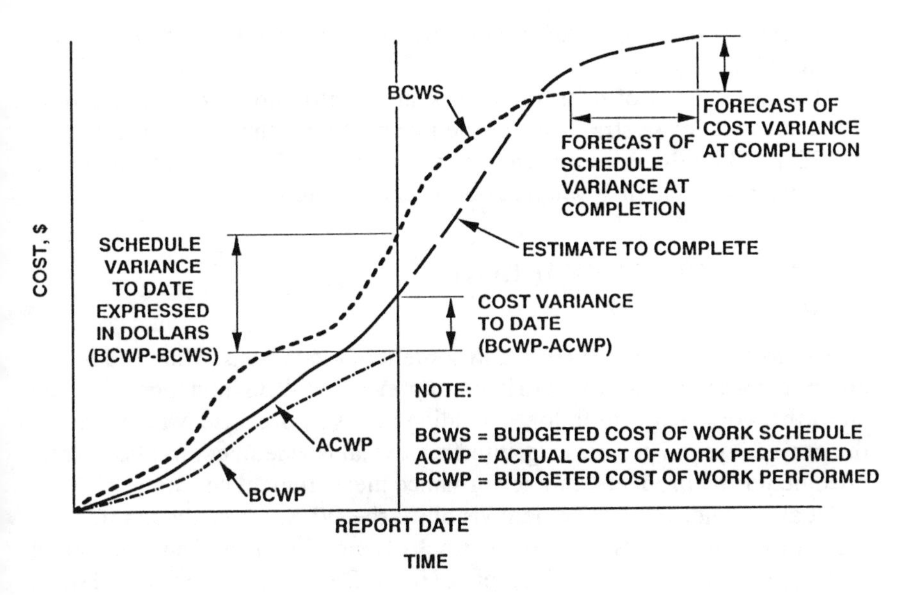


**FIGURE 15-5.** Performance measurement.

the same names that appear in a WBS. If one discovers an item that should be scheduled that is not logically contained in the WBS, then the WBS should be changed. Similarly, the WBS should also serve as a guide for items to be included in costing and performance measurement exercises.

## REVIEWS

One of the hidden values of the major life cycle reviews is that one learns a great deal about what is happening in the often intensive process of preparing for the reviews. Major reviews, at a minimum, consist of the system requirements review, the preliminary design review, and the critical design review. The nature and content of these reviews is discussed in Chapter 3. While there is generally one SRR and PDR, there can be a number of smaller CDRs in addition to the system-wide major CDR. Examples of smaller CDR's are those devoted to software alone, critical subsystems, or higher-risk hardware items. Smaller CDRs are an excellent means to benefit from the thorough critique of third parties included on the review board and to gain visibility as to your true status.

The systems engineer also has complete freedom in calling for less formal peer review meetings at any time. These include *requirements walk-throughs* and *design walkthroughs* for either hardware or software and *code walk-throughs* for software.

The lowest level of constant review in an effort to identify impending problems and devise strategies for their correction is the weekly meeting of the SDT. While the excessive calling of reviews is, of course, to be avoided, it is still a good rule that, when in doubt, call a review.

## A SOUND WORK BREAKDOWN STRUCTURE

The importance of the structure and content of the WBS is evidenced by the amount of discussion devoted to it in Chapter 5. The WBS influences virtually all of the work that is to follow, including how you organize, who works on the project, system partitioning, interfaces, what is scheduled, what is costed, documentation products, and performance measurement concepts.

Of all the aids to management visibility, the WBS can easily be rated the most important. It is also the firm bedrock of commitment and understanding for all parties involved. Development of the WBS is an early and crucial stage in the process of systems engineering. If consistency with the WBS becomes difficult because of its structure or content to the point that it does not reflect reality, be prepared to change it. While this is not desirable, it is better than abandoning it. Always maintain the WBS as a valuable living management tool.

## MARGIN IDENTIFICATION

The systems engineer is responsible for taking the lead in identifying and quantifying important system commodities that must be tracked throughout the design, development, and test phases. This responsibility includes partitioning of commodity budgets across subsystems and allocating sufficient contingency and the tracking of margin status.

Diligent assessment of margin status at formal reviews, design team meetings, walk-throughs, and informal conversation are all useful methods for maintaining constant visibility with regard to margin status.

For example, a satellite system or any flight system will almost always involve developing commodity budgets for overall mass, power, center of gravity, payload weight, and volume in addition to other items related to the mission that are deemed critical. Earth-based communications systems typically involve budgeting resources, such as overall system throughput, response times for specific message types, memory, error rates, system availability, and so on.

The setting and managing of margins entails:

1. Determining system excess capacities that are desired for the product baseline that is to be delivered to the user.
2. Determining of initial margins to be applied to achieve product baseline goals and to assure adequate margin throughout the uncertainties of design, fabrication, and test.
3. Developing a plan to allocate margins across subsystems and to periodically monitor the disappearance of allocated margins.

Note that, ideally, margins devoted to the development process should methodically dwindle to the target margin for the product baseline. The target margin may or may not be zero. The situation is very similar to planning for and using up financial contingency. If a project comes to conclusion without expenditure of planned financial reserve, the project (and the system engineer) may be subject to criticism for tieing up money. If a cost overrun is experienced that depletes planned reserve, your good friends in management might also find a reason to meet with you—and rightly so. The terms budget management (i.e., dollars) and commodity margin management are interchangeable in this sense, also for a good reason. Thus, setting system margins and their allocations across subsystems requires the same conceptual mix of aggressiveness and conservatism that financial planning does.

Typical resources that often require budgeting are:

Mass;
Power;
Computer memory;
Response time;
CPU utilization;
Bandwidth;
System availability;
System reliability;
Error rates;
Target circular error probability;
I/O ports;
Volume;
Size;
Payload;
Thermal;
Expendables;
Survivability; and
All the other "ilities."

And anything else systems engineering believes should be monitored.

## Estimating Margin Values

The values discussed below are given as guidelines only. The actual values used for both hardware and software will depend on the systems engineer's level of confidence and the assistance derived from design team members. The use of outside experts is also encouraged in this effort. It is always preferable to gather values based on experience from similar projects. If empirical data cannot be acquired, it is wise to widen margins as much as possible while still meeting system end-to-end performance requirements.

Typical hardware margin values are given in Table 15-2. The widest margin is given to guesses and "good ideas." Clearly, this is the category in which the greatest risk lies and 20 percent probably represents a lower limit. Sketches, drawings, and schematics generally enable the making of some kind of assessment as to part types, part counts, required power, circuit performance, availability, and so on. If these analyses are carried out with care and if confidence in their findings is good, a 10-percent margin should be adequate. If not, scale up. When dealing with off-the-shelf known components that are mature and have specification sheets that can be believed, or if knowledge is based on reliable experience, a 2-percent margin should be reasonable.

Occasionally, a system segment, element, or subsystem may consist of an exact replica of one previously employed. If it is felt that there is little chance that the new application may require unforeseen changes, the margin on commodities associated with the replica could be as little as 0.5 percent.

A starting point for determining software-related margins is given in Table 15-3.

Computer memory is a commodity that is almost always consumed at a rate higher than anticipated. When engaged in building a system very similar to one that has been built before, or if good documentation on a very similar system is available, one may start with 100-percent memory margin with a 25-percent margin as a product baseline goal. If there is any question at all about this, use a much bigger margin. Because memory is cheap, it requires

**TABLE 15-2 Hardware Related Margins—Guidelines**

| For | Margin |
|---|---|
| Guesses and "great ideas" | 20% |
| Sketches, drawings, and schematics | 10% |
| Off-the-shelf known components | 2% |
| Exact replicas | .5% |

**TABLE 15-3 Software-Related Margins—Guidelines**

| For | Margin |
|---|---|
| Memory | At least 100% to start, wind up with 25% unused |
| CPU utilization | Never allow utilization >70% with random arrivals |
| Response time | 10% to 100% |

*Remember*—Software requirements drive hardware requirements.

relatively little space, weight, power, and so on, using a more substantial margin is usually possible—often up to 500 percent.

Make estimates for all software involved. Estimates are generally easier for existing operating systems, compilers, utilities, and so on. The big unknowns lie in new executives and application software. Break these down as much as you can at this early stage, making estimates for each program even though the design has not really begun. Clearly, at this stage, you must err on the conservative side.

After delivery, software systems undergo software maintenance, which invariably calls for periodic operating system upgrades, general software package upgrades, corrections due to failure reports, and enhancements to application software resulting from routine change requests. Considerations for memory management should address these additional memory needs in addition to the uncertainties involved with the development cycle.

CPU utilizations must never exceed 70 percent, with random arrivals for requests for service (see Appendix A on "Elements of Queueing Theory"). This means that, if a 100-percent margin is set at the outset, the initial design will call for a 35-percent CPU utilization. CPU utilization is a very important commodity to plan for and monitor. There are a lot of systems in service that run too slow under peak average loads or that meet requirements at installation time but rapidly deteriorate with reasonable growths over the system's lifetime.

Meeting end-to-end system response time goals also typically requires applying conservative margins. Again, the margin assigned depends solely on the level of faith that requirements can be met. Should there be any serious reservations, try to start out with a 100-percent margin.

In dealing with computers, it is almost always best to begin by setting software margins. Stated in alternate terms, computer hardware requirements should be driven by computer software requirements. That is, setting computer software margins is the starting point for setting computer hardware margins. There are rare cases where it can happen the other way around, such as instances where you might be constrained to the use of inherited hardware for an upgraded or new system, but it is seldom desirable to start this way.

Many military systems call for preplanned product improvements—the so called $P^3I$. These product enhancements are improvements that are planned for as a part of the present design and development process for anticipated improved versions of the product to be implemented or added on at a later time. These considerations, in effect, call for placement of higher levels on a pertinent set of your margins. This, in itself, is no small exercise and must be considered in these early stages of setting margin values.

Actual estimates will, of course, vary around these guidelines. The margin values selected are a direct consequence of the level of security associated with the desired targets. Ideally, margins are determined from solid experience. When this is not possible, the expertise on the design team should be called upon to perform analysis, simulation, or even preliminary breadboarding or brassboarding. Margin setting is also a valid issue for the SDT to call upon outside experts as required.

## Margin Management

Margin management is the process of tracking the methodic disappearance of any excess in margins that have been allocated across subsystems and the systems engineer's own system margin reserve. For example, Figure 15-6 depicts a hypothetical distribution of a margin allotment over four subsystems. Suppose the goal is to realize a system response time of 100 milliseconds. Based on analysis and consultation with the subsystem COGEs, the systems engineer might allocate 30ms, 12ms, 18ms, and 20ms across the four subsystems, adding up to 80ms. The reserve of 20ms to make up the 100ms final target is a measure of the systems engineer's confidence that subsystem goals are realistic. Of course, the COGEs are aware of the 20-percent margin that the design team is holding out, and it must be impressed upon all that this margin is not readily available for depletion, especially early on. Thus, it is important that their estimates going in are based on the best data available and not upon the assumption that there is plenty of "pad" available. It will not be atypical, as the design matures, that each of the COGEs will want the entire 20 percent for themselves.

Margin is always easier to give up as the design becomes more mature and performance characteristics are better understood. Conversely, it is much more dangerous to give it up early in the process when uncertainties still exist. Further, the uncertainties faced are generally not distributed evenly over all subsystems, but are concentrated in one or two of the subsystems. Impress

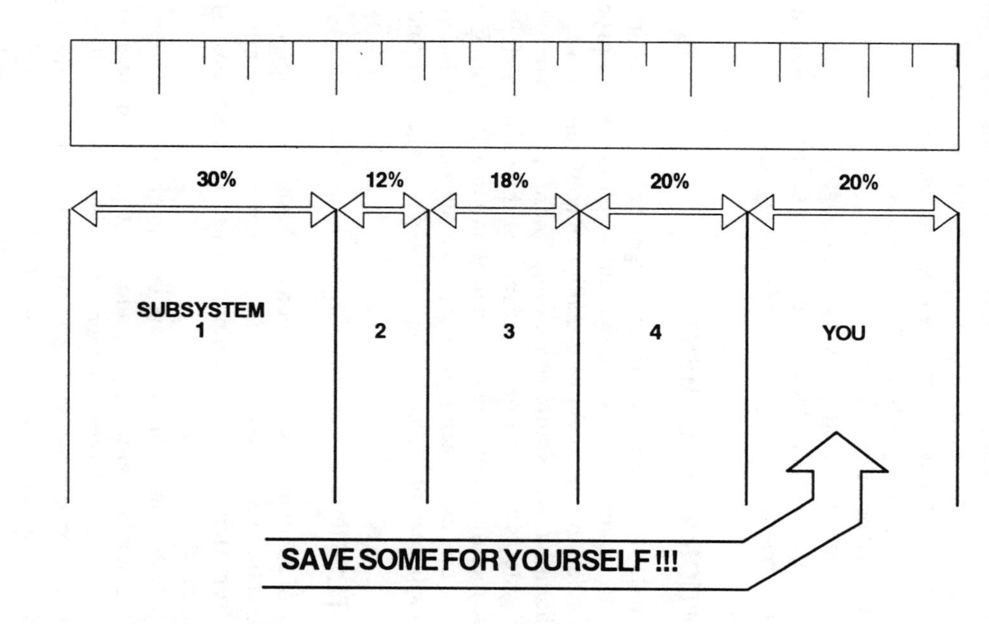


**FIGURE 15-6.** Allocate margins over subsystems.

upon the team, as you go through margin planning, that, if a COGE suddenly requires a part of the withheld system margin 1 week into the design process, he or she is not likely to get it that easily and that a hard revisit to the subsystem design approach will be called for. Also, look for ways for the subsystems to trade among themselves. Don't give up reserves early in the process without a struggle.

The design team is the major forum for the initial setting of margins and for the management of their organized dissipation. The weekly design team meeting is the best resource for maintaining and constantly monitoring margins. This is done by simply asking what effects on margins are anticipated as design concepts and possible changes are discussed. If doubts arise, use the team to carry out any required analyses. All design reviews and design walk-throughs should always include an up-to-date formal commitment on the status of all margins.

## CONFIGURATION MANAGEMENT

The principles and value of configuration management are covered in Chapter 8. Maintaining control over the design and implementation process is absolutely dependent on the effective execution of configuration identification, control, auditing, and status accounting. These ingredients must be tailored to the systems engineering paradigm that the project is following. Achieving these modifications may require negotiation with existing institutional policies. In any case, each ingredient must always be included. Any failure to do so will inevitably result in rapid loss of management control.

## ADEQUATE TESTING

The importance of adequate testing has been emphasized in Chapter 14. The cost of error detection and correction during system development increases exponentially, from the point of preliminary design to operations. The increase can be as much as 100 percent.

The ability to detect and correct errors in a timely manner is largely a function of the systems engineering development paradigm used. In the staircase models, the single largest source of errors in system development is in the production of poor requirements and specifications. For this reason, when using the staircase approach, the importance of using a single system requirements review and preliminary design review is emphasized. In these models, the cost and schedule impacts of excessive errors is generally not understood until system integration and test takes place.

The early prototyping and spiral models greatly reduce errors in the early stages of systems engineering. The rapid development model provides the greatest ability to uncover errors, correct them, and, if need be, alter course prior to final system acceptance.

For any of the paradigms, the astute use of early testing strategies can significantly enhance the visibility that is so dearly sought by the systems engineer. This includes utilizing aggressive low-level testing on both hardware parts, subassemblies and components, and software procedures, modules, and CPIs. Particular attention must be paid to creative lower-level interface testing to the extent that it is realistic.

Investing resources in the early detection and correction of errors is a wise idea for at least two reasons. First, early testing is a sound tool for technical visibility. Second, every error found early directly translates into a savings of time and money later.

## PERSONAL SKILLS

There is no question that the systems engineer must possess sufficient technical skills to communicate with a large variety of specialized experts. Possession of technical knowledge is a major factor in the ability of the systems engineer to gain the respect of his or her colleagues and SDT members. The gaining of this respect for the ability to handle technical issues through time is an absolute necessity and must be patiently but thoroughly earned.

While technical expertise is a necessary condition for success, it is not a sufficient condition. In any interdisciplinary setting, the ability to deal with the diversity of issues related to human relations alone is also imperative. The keystone to developing and improving our human relations is communications.

The human communications channel is extremely complex. This complexity is simplistically depicted in Figure 15-7. The originator of an attempt to communicate starts with an idea, shown as Idea 1 in the figure, that is presumably worthy of communication. The idea is an abstraction and must first pass through what is termed the chopping block of language. The outcome of this transform is a filtration and is dependent upon such factors as the originator's language skills, sensitivities, attitudes toward the listener, and current set. Psychologists refer to the set of an individual as being comprised of the sum total of that individual's history strongly influenced by recent events. Your set is the composite of your attitude, knowledge, and recent experience at a given point in time, which largely determines your behavior. It includes, among other things, the probability of your choice of particular words.

After the chosen sentence or sentences are uttered, the listener (if not speaking at the same time) processes the information. We assume that both parties are in the same room, with a tolerable noise level, so that the

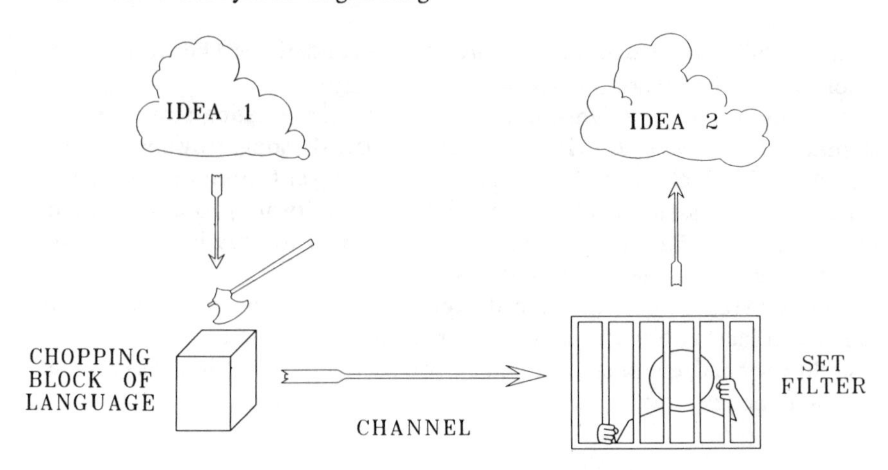


**FIGURE 15-7.** The challenge of communications.

channel is basically clean. The received string is then transformed through the listener's set filter into a second idea. The probability of Idea 2 being anything like Idea 1 is highly dependent on all of the factors mentioned above, as well as on the subject matter and how open the listener's mind is to the offering.

Technical concepts are generally easier to communicate than highly abstract notions because the language tends to be more structured and definitions tend to be less fuzzy. It can still, however, be a very difficult process because even the most acceptable technical definitions are not always universally stored in everyone's memory in the exact same way. Typically, they are not.

The problem is greatly compounded by considering the attitude of the listener toward the communicator. It has been professed that, at any point in time, each of us is at one of four mental levels with each individual we encounter [2].

The lowest level, and the least desirable, is the *closed mind.* At this level, the listener's attitude is basically one of avoidance, which can be summarized as the "You again?—Get lost and don't come back" attitude. Nothing can be said or done at this level to gain serious attention. Don't persist in trying, and don't offer criticism, as your cause cannot be advanced.

The second level is the *open mind.* At this level, party 2 will actually listen to party 1, but is reluctant to believe what is said without substantial outside corroboration. The I'm-from-Missouri behavior can be summarized as the "Sound's kind of funny—let's see what Charlie has to say about that" attitude. It is not an entirely unhealthy attitude. People at this level will "look it up" themselves or seek third-party concurrence with what you say.

The third level is the level of *confidence*. This is an extremely productive level, where participants have confidence in their colleagues. Conversations are not confrontational, but only seek to achieve clarification and understanding. The communication goals are oriented toward problem solving and not toward personalities. At this level, we hear sentences like, "I didn't follow you on that" or "what is the basis for that?" The level is characterized by the "say something to convince me" attitude. Significantly, at this level, party 2 does not immediately feel a need for outside corroboration. There is a fundamental expectation that party 1 has sufficient knowledge to address the issue at hand.

The fourth level is nice sometimes, but can be dangerous. It is the total *belief* level characterized by the "whatever you say is absolutely right" attitude—"it's OK by me."

The level or mix of levels that one experiences is not only a function of who one is speaking with, but can change with the same person depending on the subject matter—and can alter through time for a given subject matter as relationships grow or change.

While this scheme may be somewhat rudimentary, it is still useful in ascertaining the probability of success in effectively communicating with a listener. The concept is also useful in recognizing how to conduct yourself in given situations and in clarifying what course of action must be taken to improve the situation.

If you are party 1 and perceive that you are on the closed mind or open mind level with a party 2, you may concentrate on determining how to build toward the confidence level with regard to the particular issue at hand. In many cases, it is simply a matter of building your own technical skills. It may also involve improving interpersonal skills.

When confronted with a level 1 relationship in private life, the option to simply avoid the party exists. In the work environment, it is desirable to have the party removed from the team. If this is not possible, the problem must be worked in a positive way. Initial efforts should concentrate solely on advancing the party to level 2. While this may not be a trivial exercise, it can be first approached by simply agreeing with and complimenting the person to every extent possible, even when what is said irks you a little bit. In particular, if the conversation revolves around a topic that is not important to your goals as systems engineer, there is simply no point in disagreeing with a party at level 1. It is best to initiate the "agreeable"approach in private conversations that you seek out and, of course, at every opportunity that presents itself in team meetings.

"Mmmm—that's an interesting point," you say, "but what about such and such?" You will then be told about such and such. After you are told about such and such, you say, "I see. I never would of thought of it that way. Interesting."

It is a skillful path to trod upon, as it involves acknowledging that a particular point is worthy of consideration without specifically committing to its implementation. Negative, closed minded people are generally not liked, and it is a peculiarity of their makeup that, if you agree with them often enough, you are likely to be a rarity among their circle of communicants. They may eventually begin to actually listen to you—that is, transform to level 2.

Moving from level 2 to level 3 is a lot easier in the workplace, as it can almost always be accomplished by improving your technical skills in the arena of the parties' expertise. Again, this takes study and patience, but once the homework is done , the subtle dropping of a well-chosen technical term at the right time can dramatically affect one's opinion of your capability. In the end, however, you can't fake moving from level 2 to level 3. You must earn it. But at least the strategy is clear.

In a continuing effort to build the confidence of others and to gain the support that is constantly needed, it should be useful to periodically review the following points:

- Take time out to sharpen your technical skills.
- In conversation, separate the important from the unimportant. If it is not important, avoid confrontation. People remember confrontation. You don't need confrontation.
- Take every opportunity to support your workers and team members —at your meetings and particularly before upper management.
- Always do three things: Listen, Listen, and Listen.
- When you have an opportunity to give in on an issue and still meet your goals—do it. Amass tokens.
- Care about the success and careers of your team members. If the team succeeds, you succeed.
- Don't pretend to know it all—you don't.
- Avoid negative rejection. If you must reject a position, try to be positive in your rejection. Most bad ideas are rejected by team members after you ask the right questions. The Socratic method can be powerful in a team environment.

Alternatively, if you are a systems engineer and perceive that you have a total belief relationship with an important team colleague, you may from time to time consider dropping them to the confidence level, if for no other reason than your own safety. Nobody is always right.

The systems engineer should constantly strive to assess the levels of human relationships encountered with all team members, as well as management, and use this information to develop specific case-by-case strategies for building toward professional relationships at the confidence level.

## Vector Management

Projects begin by conforming to top-level plans, such as project plans and systems engineering management plans. The larger plans give rise to more detailed plans, but the big plans are always aimed at a goal, namely the successful delivery of a system at some point in the future. More often than not, specific near-term goals change as implementation proceeds due to changing requirements, issuance of waivers, or design impacts that could not be foreseen at the outset. Change is expected, and the capacity to respond positively is essential. Ideally, the big plan does not change so drastically as to jeopardize the ultimate achievement of a system that is acceptably close to the original goal.

In this milieu, the big plan can be likened to a large vector aimed in advance to intersect the goal. The vector has momentum analogous to the vector that sets the main course of biological evolution in a given environment. It too has momentum, but is modified slowly through time by the ever-present changing gene pool. The gene pool is a little vector that slowly modifies the direction of the big vector.

The analogy to systems engineering is represented in Figure 15-8. It is the responsibility of the systems engineer, along with the design team, to assess the current location of the goal and the integrity of the original plan and to technically modify the direction of the vector that represents the big plan, such that it steers home to the target.

This analogy offers a useful top-level visualization for the systems engineer. It suggests that, from time to time, the apparent direction of the SDT can seem to change dramatically in an effort to steer the big vector home. The systems engineer must be willing to change a viewpoint with regard to requirements, design approaches, margin management strategies, and so on, when it is clear that such action should be taken. The vector management concept not only allows such change to take place, but also justifies it in the more global context. There can be a thin line between being perceived as resilient and being perceived as constantly vacillating. While constant change is not desirable, the systems engineer must not be reluctant to change through fear of criticism when change is indeed warranted. Constructive change is a positive act. The vector management construct inherently includes such capacity for change because the systems engineer does not control the big vector, only the little one.

Once a manager of mine, after sitting in on my design team meetings for

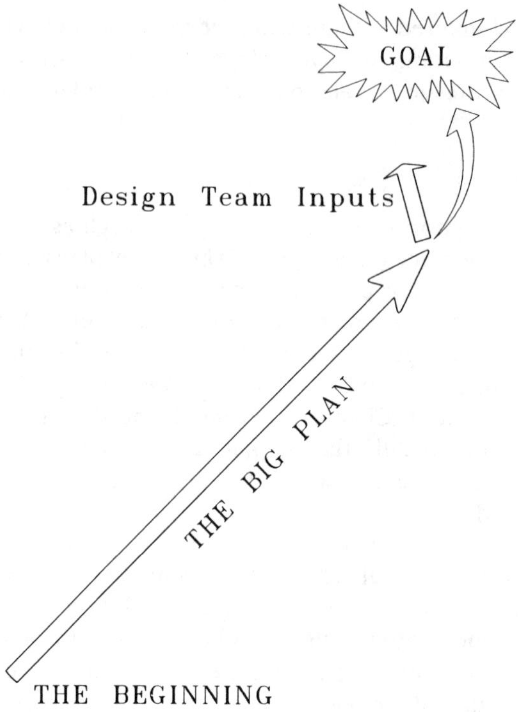

**FIGURE 15-8.** Vector management.

two consecutive weeks, asked me, "How come two weeks ago you made the exact opposite schedule change that you made this week? What direction are you going in, anyway?"

I pulled out my vector management picture and told him that I was still taking the big vector home. He understood that, in the context of two weeks, the change appeared to be a 180-degree switch, but in the overall context it was a small reactive and positive readjustment. There were no more questions. A few months later, I heard him use the same concept in a presentation.

## ORGANIZATION

Organization has a profound effect upon the ability of programmatic managers and technical managers to maintain visibility. Although seldom admitted, organizational structures alone are frequently among the major reasons for system failures. The role of the systems engineer is an extremely delicate one. Before accepting such a position, the aspirant should be critically sagacious with respect to the impacts that organizational structure can have on the ability of the systems engineering role to succeed or fail.

There are two generic types of organizational structures—the *functional organization* and the *matrix organization.*

The functional organization, depicted in Figure 15-9, is organized around projects. Each project structure repeats the necessary functions to achieve project goals. Figure 15-9 gives examples of what these functions might be. The particular functions are determined by the mission of the organization in government or in the marketplace.

The basic matrix organization is portrayed in Figure 15-10. The matrix structure consists of a pool of separate technical capabilities. The pool is tapped by projects as they come and go. Project managers decide on who is needed and call upon the technical resources to provide the appropriate personnel from across the matrix.

The functional organization is basically fashioned for situations where products or services remain fairly stable over time, such as the automobile, steel, or home entertainment industries. Matrix organizations are designed to deal with varying products or services, such as research or high-tech enterprises, where project or task work is predisposed to change over time. This is not a rigid rule, as the reverse may occur in given instances and mixes of the two are also in evidence.

The functional organization tends to have more stability with regard to both project structure and personnel turnover from project to project. It also fosters the creation of project turf.

The matrix concept is inclined to exhibit less conformity with regard to project structures because project managers tend to have more freedom in the way they organize in adapting to projects of different natures. Also, it is

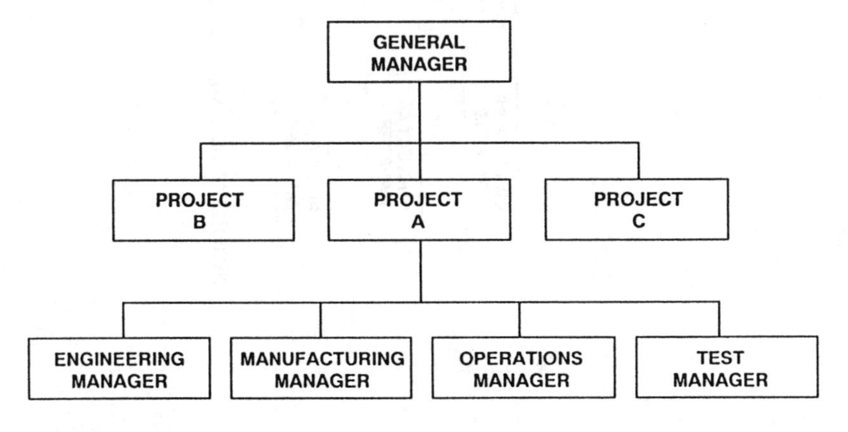


**FIGURE 15-9.** Functional organization.

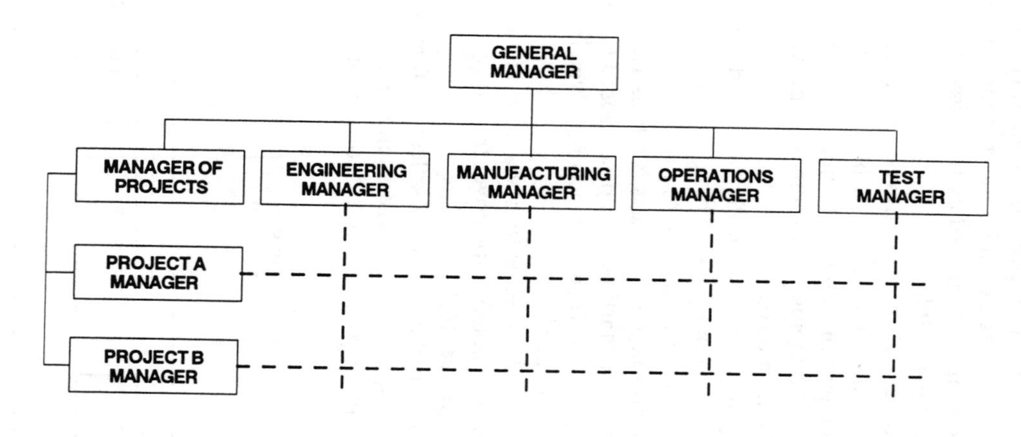


**FIGURE 15-10.** Matrix organization.

generally easier for personnel to move from project to project in a matrix organization. The matrix organization fosters the creation of discipline-oriented turf.

In reality, implementing these generic structures involves a great diversity of expression. What follows is a description of four distinct, actual project organizations, each of which was involved in a costly system failure. The failures are directly attributable to the inability to perform the systems engineering role due to structural flaws alone. The examples are taken from both functional and matrix organizations. A single critical flaw across all examples is then identified, and a project structure designed to avoid this problem is suggested.

## Case 1

The Case 1 organizational structure is given in Figure 15-11. The structure was devised by a project manager in a matrix organization. A first observation on the arrangement is that it is not clear who is in charge. Systems engineering, software management, environment & QA, and resource management all appear as stem winders to project management. In this design, the path to authority for a subsystem COGE was not well defined. In practice, when a COGE identified an issue perceived to have system impacts, he or she first decided whether it was primarily a programmatic issue, a software issue, or a systems engineering issue. The COGE would then approach what was considered to be the appropriate person. Unilateral decisions were often made that affected all functions without the timely inclusion of counterparts. The concept of the SDT being the focal point for handling all issues was slowly eroded.

Further, the chart confuses the relationship between the systems engineer and the software manager. The systems engineer has technical responsibility for the complete system, which includes both hardware and software. The indefinite structure eventually led to confusion as to who was technically responsible. The confusion, in turn, gave rise to discussions among the workers, who were trying to determine who had what responsibilities—an issue never clearly addressed by project management. The manager's perception was that he was strapped with a pack of power hungry animals.

It should also be noted that the structure does not conform to the generic WBS for the following reasons:

1. Environment and QA are two different things. While it may be reasonable to include QA as a staff function, environmental issues should be handled through specialty engineering representation on the SDT, as these matters directly effect the system design.

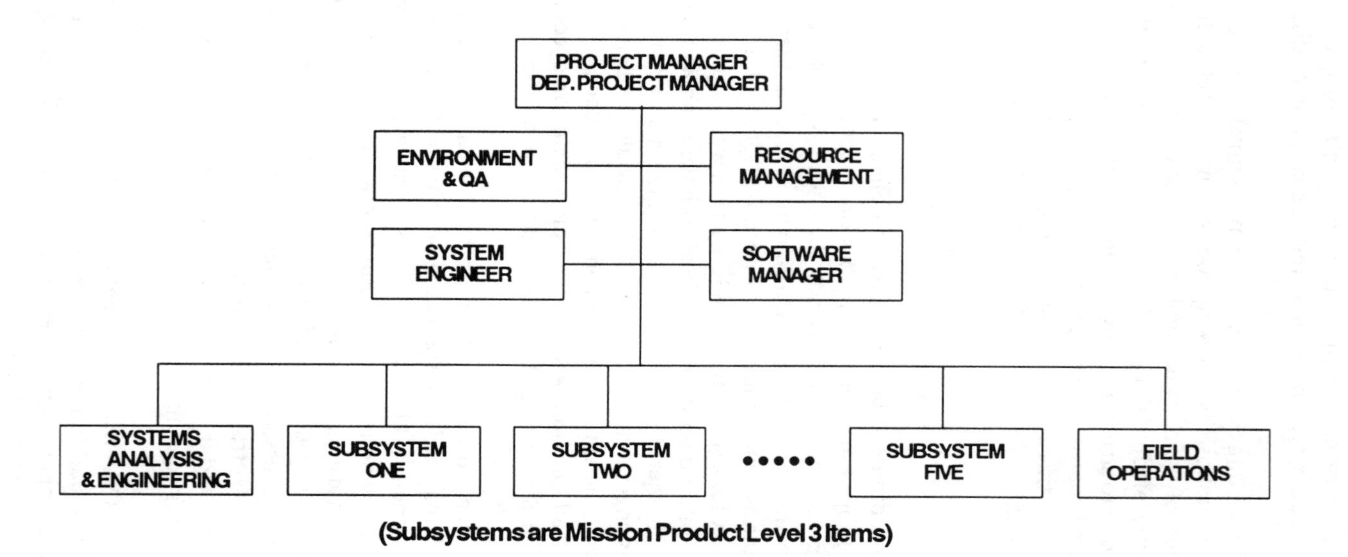


**FIGURE 15-11.** What's wrong with this organization?

2. There is no clear indication of how testing functions are to be accomplished.
3. The closest function to system-integrated logistics support is the box entitled Field Operations. There is no clear provision for the remaining ILS items of personnel and training, technical data packages, transportation and handling, facilities, or support equipment to sustain those operations.

The role of systems engineering was substantially hindered on this project to the point that the project was unilaterally canceled by the sponsor when it was evident that a cohesive system design was not emerging.

## Case 2

System development was originated for a software-intensive system to be used for strategic planning of resources at the national level. The project initiator, who was in charge of resource planning, existed in a separate department from the three departments in which the resources actually existed—Departments A, B, and C. The problem was viewed as a software problem, and a contractor was selected and brought in. The contractor did exactly what the project manager asked: he developed a software system.

The project manager knew exactly what data inputs and outputs were needed to solve *his* problem. It was assumed that, when the system was developed, Departments A, B, and C would want to use it. There was no design team formed—hence, no real user interaction. Figure 15-12 summarizes the condition.

The system was completed but never used because the project manager made two very serious errors:

1. The project manager did not organize the effort around a SDT to include all the system users. It turned out that the required data that presumably existed in Departments A, B, and C were not organized in the form required by the software design. Further, the department head of A, B, and C, having not been involved, simply found ways not to provide it.
2. The project manager did not understand that it is not sufficient simply to allocate resources. In resource allocation settings, conflicts arise and the system must also include a mechanism for resolving conflicts and subsequent replanning on a case-by-case basis. The system should have involved the establishment of resource conflict resolution teams and structured inter-department operational human interfaces. A proper determination

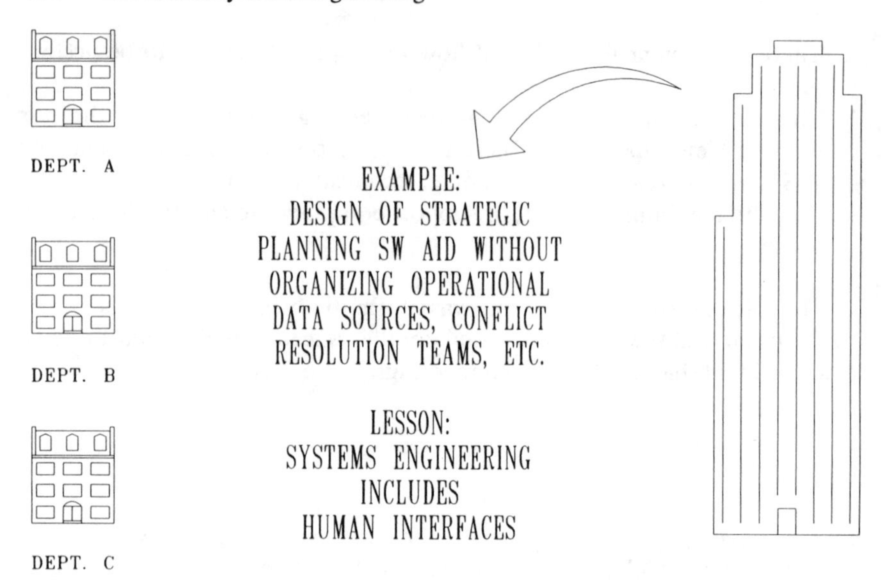


**FIGURE 15-12.** Organizations are not always structured to use your product.

of user requirements would have determined this. The software was only a small part of the system—an automated aid to support the larger system mission, which was never considered.

In short, the system was never used because the project manager did not understand the "system" from day one—hence, he could not organize properly to solve the problem.

### Case 3

Case 3 adhered to the classic staircase with feedback approach. The paradigm was followed faithfully with the development of requirements and specifications and the holding of all proper reviews, and so on. All the right steps were executed, but by the wrong people. Because the project was so "important," all the communication between the two organizations was done at upper management levels. Top management knew the generic process, but was unable to empower responsibility.

The real users existed at the bottom of the user organization, and the real implementors at the bottom of the contractor organization. They communicated with each other through a number of levels of management in each organization. The communication path is depicted in Figure 15-13. There was

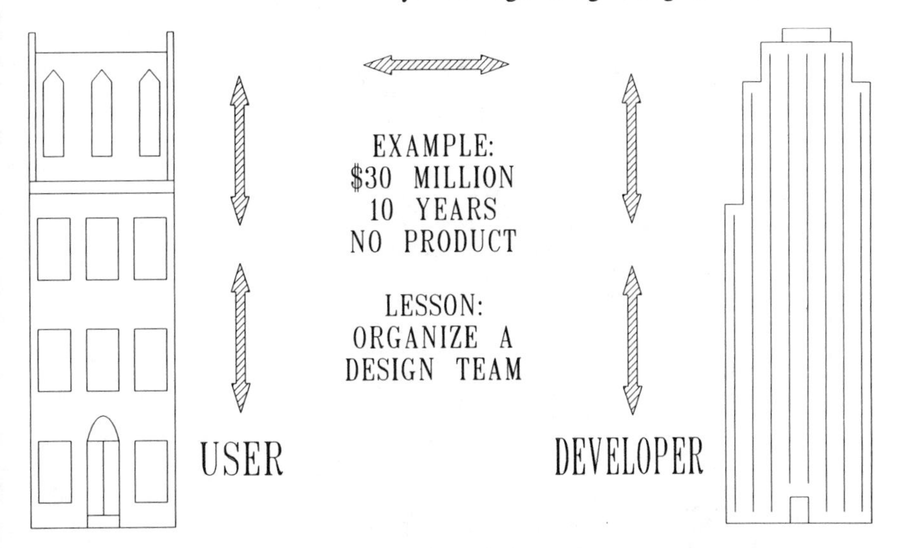


**FIGURE 15-13.** Organizations can destroy communications.

no nitty-gritty weekly SDT structure in which the actual users and developers could interact. They didn't even know each others' names.

After 10 years of effort and an expenditure of $30 million, the system did not work properly upon delivery. Today, they are still "working" on it.

## Case 4

Case 4 involves another matrix organization. The organization is very large, as are each of the discipline-oriented members of the matrix. The major roles of the matrix members are somewhat defined; however, they are dispersed geographically so that each of the members has found it necessary over time to implement overlapping capabilities at their own sites. One of these overlapping capabilities is systems engineering. The title appears everywhere.

Top management is also geographically isolated, but what is more significant is that it exercises programmatic control only over the matrix members. Technical authority over day-to-day events is distributed—delegated to each center of expertise. New programs are assigned by top management to a matrix member or, in the case of big projects, to a group of members. Matrix members may also call upon each other for support (see Figure 15-14). The success rate has been higher when the assignment has been made to a single member or when support is agreeably solicited among members. In fact, the members work fairly well by themselves.

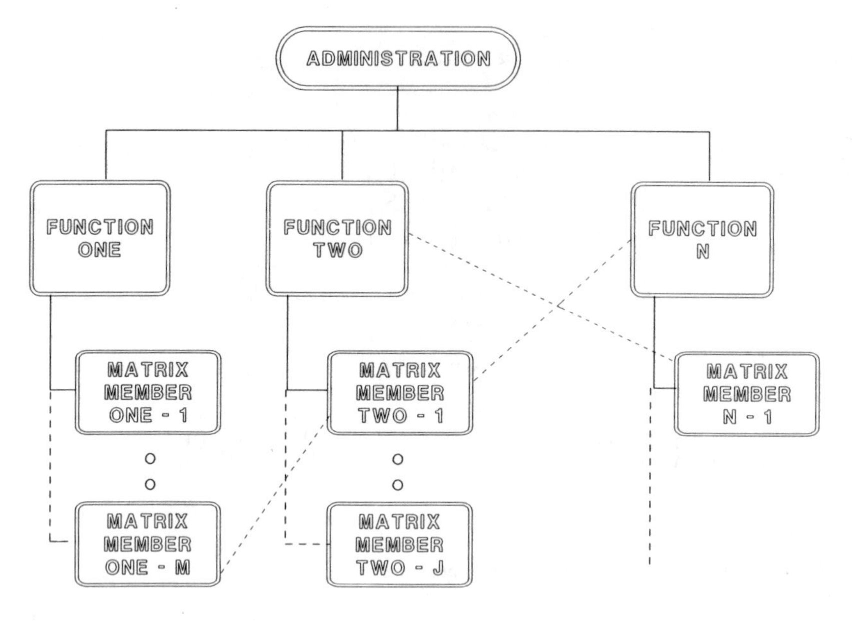


**FIGURE 15-14.** What's wrong with this organization?

When a large project is assigned to multiple members by top management, turf issues typically arise. This is because the members exhibit a common characteristic of matrix participation: each one thinks they're equal, and some of them think they're better. When such an organizational structure assigns the role of systems engineering to one of the members, the result is that no one is really in charge. There is no central day-to-day technical component with the authority to manage all the members. The structure simply does not provide the authority for systems engineering to take place under these conditions. The responsibility exists, but not the authority. The result is that big multi-member projects are placed in serious jeopardy from day one, due solely to organizational deficiencies.

These four cases may appear extreme. Unfortunately, they are not only real, but involve some of our nation's most prestigious private engineering and government organizations. What is more impressive is that in each case the project and organization managers found reasons for their system failures that were totally unrelated to their organizational structures and management styles. The inward look is, evidently, extremely difficult. These structures and ones like them are created over and over again and continue to be sustained, even by managers that confidently declare, "I would never make a mistake like that."

Note that each of these cases has a common thread: none of them were structured to allow the generic systems engineering process to properly flourish. All of them contain significant turf barriers. The most perilous consequence is that SDTs and user interactions were seriously compromised.

The organizational issue has come under increased attention with the advent of concurrent engineering and modern quality concepts. In the conventional nonconcurrent approach to development, each organizational entity tends to determine requirements in a relatively isolated environment. Each has their own turf. Thus, the functions of developing user needs, designs, manufacturing needs, testing, marketing strategies, and operational support have typically lacked vital degrees of coordination. Each entity would basically pass results of their work "over the wall" to the next. Further, suppliers are typically sought out after the design process.

The integrated product development concepts of concurrent engineering call for total coordination of all of these functions, independent of the development paradigm in use. Complete consideration of all aspects that influence product development also includes design integration with suppliers.

Organizations initially undertaking concurrent engineering often begin by organizing design and manufacturing functions under one new department. This is probably the case because design and manufacturing have traditionally been areas of considerable contention. Designs that have not adequately considered producibility issues are often bounced back and forth incessently until time runs out and management simply orders a "go ahead."

While such reorganization is a beginning, it addresses only a part of the overall need. Concurrent engineering can best be achieved when all WBS items are organized under the systems engineer, integrated product development teams are organized, led, and coordinated by the SDT, and top management totally supports the authority of that organizational structure.

The lesson for the systems engineer is simple. Don't take systems engineering assignments in an organization that is not structured to do systems engineering. You must be empowered to run your show in a manner consistent with the generic principles of systems engineering. In short, if you are serious about being or becoming a systems engineer:

*Don't take responsibility without authority.*

If you are not given sufficient authority in the systems engineering role, you will not be able to effectively implement or control the processes and tools of systems engineering that are required to acquire and maintain technical visibility.

A project structure that is more amenable to execution of sound systems engineering principles is shown in Figure 15-15. Note the following characteristics:

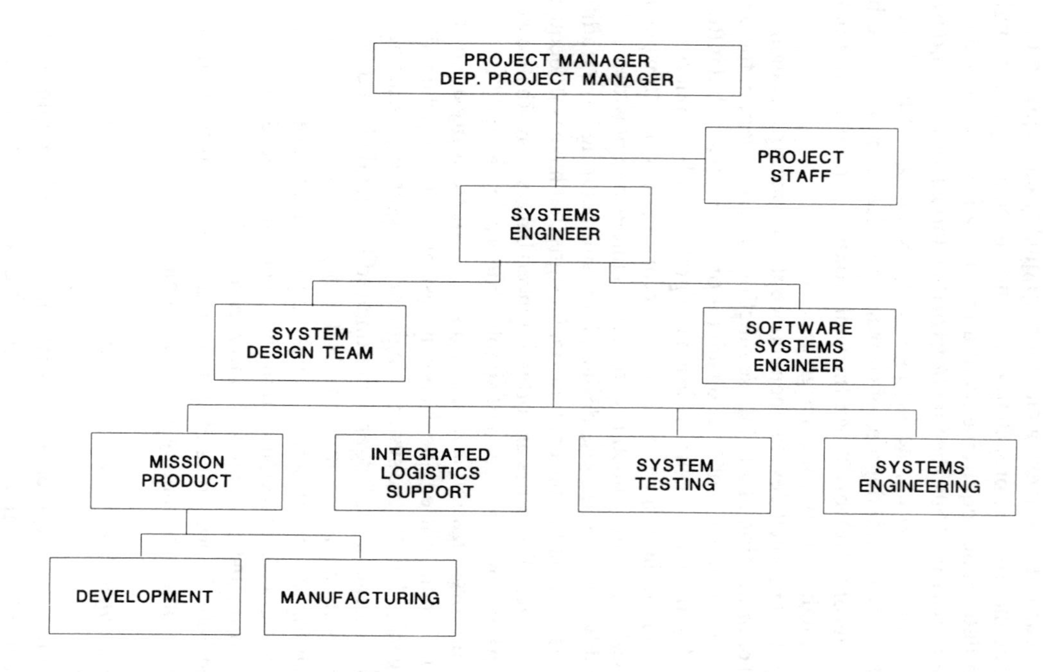


**FIGURE 15-15.** A generic project structure.

1. The systems engineering function is clearly in line to project management. While project management is ultimately responsible for everything, the PM office is primarily concerned with programmatic and upper management interface issues on a day-to-day basis. Complete technical responsibility lies with the systems engineering function. The authority and empowerment is clear.
2. Software systems engineering clearly reports to systems engineering. It is also superior to all of the four major functions at lower levels. This is because total system software engineering not only involves operational software for the end mission product, but also encompasses software issues related to ILS (including training and maintenance) and testing. The mission product software is often integrated with these other software needs. The software systems engineer must be given the authority to address the total system software picture, not just the required operational software.
3. The SDT is clearly delineated and is superior to the four basic lower-level elements. The SDT is clearly organized and managed by the systems engineer. It is also the clear focal point for executing concurrent engineering should these concepts be employed.
4. The four basic generic WBS components of the mission product (subsystems), ILS, system testing, and systems engineering staff are clearly represented. For projects where production is involved, the mission product would be further broken down into development and manufacturing WBS items.

An organization such as this may not guarantee success, but it will provide a structure that, with proper technical and programmatic management, should not inherently increase the probability of failure.

## THE SYSTEM DESIGN TEAM

It is appropriate to close this treatment of systems engineering with final comments regarding the SDT. The SDT is the nucleus for executing all aspects of systems engineering. It is unquestionably the major instrument that guides system development and through which the systems engineer maintains control and visibility.

Systems engineering should never seriously be attempted without a formal SDT. But simply establishing a team is not sufficient. Management must also understand and support the need for the team. This support should be evidenced by the organizational delegation of complete responsibility and authority to the systems engineer for all system-related issues and activities. Further, the systems engineer must be given the ability to structure the SDT to effectively perform all systems engineering functions (including concurrent engineering) and to include the key membership roles covered in Chapter 7.

Finally, the systems engineer must be given the freedom to authoritatively adapt, modify, and implement those generic processes and procedures covered throughout this book. But don't stray too far!

With these conditions met as a minimum, the stage will be set for the systems engineer to exercise the technical and interpersonal skills required to attain a convincing and productive leadership role. The role is a demanding one. It is also a delicate one that must be balanced with leadership and respect for others, marked by the establishment of consistent rules and the frequent, but sincere, dispersing of credit to your team.

If it's going to happen for you, it's going to happen through your SDT.

**References**

1. Machol, Robert E, Ed. 1965. *System Engineering Handbook.* New York: McGraw-Hill Book Company.
2. Reilly, William J. 1982. *Successful Human Relations.* New York: Harper & Row.

# Appendix A

# Elements of Queueing Theory

A knowledge of the principles of queueing theory is a significant asset for an engineer involved with systems where waiting for service of any kind takes place. Queueing theory, along with Reliability, Availability and Maintainability (RAM) concepts, provide two fundamental tools useful to the systems engineer in rapidly assessing the viability of system architectural alternatives.

The purpose of this appendix is to provide an introduction to the power and utility of queueing theory to the systems engineer. The material presented is not intended to be an exhaustive treatment of the subject. Deeper and more rigorous treatments, along with derivations, can be found in the literature [1], [2].

What is presented are the basic equations that are of use in the analytic modeling of single-server and multiple-server queues—a feeling for their power by way of example and a warning regarding some of the assumptions that underlie the equation derivations.

If you are familiar with queueing theory, you may wish to simply scan this material and move on to other sections of this book. If you are a systems engineer and are not familiar with queueing theory, then you should be. While this chapter will not create experts, it will heighten the awareness of an important tool and provide sufficient background regarding the basics to facilitate discussions with those more knowledgeable in the field upon whom you may call. In any case, it is a tool that the systems engineer should definitely have in his or her arsenal.

## SOME DEFINITIONS

When people wait in line at a checkout counter or when messages wait to be processed in a communications system (or entities wait in any like structure for service of some type), they are said to be in a queue. In the queues we will be talking about, the entities that desire service wait for that service in a single line. When an entity reaches the head of the line, it encounters the server and receives service. The queue includes both the entities waiting for service as well as the single entity that is receiving the service waited for. Thus, if you are either in a line waiting for service or you are the one receiving service, you are in the queue.

A queueing system is characterized by three basic statistical considerations. These entail an understanding of the customer source, the arrival rate statistics of customers seeking service, and the service time distribution function which describes the statistics by which customers are served.

From a mathematical standpoint, the source of customers for a queue can be both infinite or finite. Finite sources are, of course, more realistic. Unfortunately, the mathematics for infinite sources is considerably simpler. Therefore, if the source of customers is finite but large in comparison to the number of servers, we can assume that the arrival rate statistics are similar to those predicted for infinite sources. In most practical queueing systems, this assumption is reasonable.

The most useful and interesting arrival rate pattern is referred to as the random arrival rate. When arrival rates are random, the inter-arrival times are exponentially distributed. That is, the probability that the inter-arrival time, *ta,* for a customer is less than or equal to some value $T$ is given by

$$P\,(ta \leq T) = 1 - e^{-T/E(ta)} \qquad \text{(A-1)}$$

where $E(ta)$ is the mean value of *ta*.

This pattern is also referred to in the literature as the Poisson arrival process. The assumption of randomness of arrivals is a fairly safe one to make when dealing with most communications and interactive computer systems as well as with most systems with human servers. This is true despite the fact that average rates per unit time may vary considerably as a function of the time of day, month, or year in many systems. Systems are typically designed, however, to accommodate those periods in which their transaction loads are highest.

In computer systems, the design is usually driven by consideration of peak arrival rates. In systems with human servers, the system design must also be capable of handling peak arrival rates; however, the staffing of that design will usually vary as a function of shifts. Thus, a supermarket may be designed to provide, say, seven checkout stands, but typically has a lesser number actually in service during off-peak traffic hours.

Another common arrival rate pattern that is of interest is one that is constant or very nearly constant. This pattern is typically encountered when items move along an automated production line and the service times at each station are also nearly constant.

The most common service time distribution function of interest in queueing problems is also exponentially distributed. This continuous distribution function, $F(t)$, is given by

$$F(t) = 1 - e^{-t/E(ts)} \quad \text{(A-2)}$$

where $E(ts)$ is the mean service time and $t$ is greater than zero.

In the queues that we will deal with, the service times required by customers are assumed to be independent and identically distributed random variables. The notation provided in Table A-1 will be used for the discussion that follows.

## GATHERING DATA

A knowledge or estimate of three basic statistical elements is required to carry out queueing analysis. They are the mean value of the service times, $E(ts)$, the standard deviation of the service times, $S(ts)$, and the average arrival rate of items to be served, $E(a)$.

### The System Mean Service Time

The first step in performing queueing analysis is to acquire an estimate of the mean and the standard deviation of a statistically significant population of service times. A customer's service time, *ts,* is a random variable characterizing the service time for a customer.

If a human being is the server, such as a dentist, a bank teller, a shopping clerk, or a radio dispatcher, the service times can be determined by measuring the service times of individual service transactions with a stopwatch and recording the data on a data sheet. If computer service times are to be measured, such as CPU or disk service times, one will generally need to develop a specific program to time tag the beginning and end times of each of the transactions of interest. In some computer systems, this data is already made available through or can be derived from existing utility programs that gather specific statistics on usage of system resources. The computer manufacturer, however, seldom knows what use is to be made of a given machine. Thus, statistical packages often provide service time measurements lumped into categories that are not appropriate for a particular problem of interest

**TABLE A-1 Queueing Notation Used in this Chapter**

| Symbol | | Definition |
|---|---|---|
| $E(a)$ | = | the average number of arrivals for service per unit time (i.e., the average arrival rate). The reciprocal of the average arrival rate is the average inter-arrival time. |
| $E$ | = | the average traffic intensity equal to the average arrival rate times the average service time ($a \times ts$). The unit of measure of traffic intensity is the Erlang. |
| $E(x)$ | = | the mean value, or the average value, of the random variable $x$. |
| $E(nq)$ | = | the average number of customers waiting in the queue, including the one being served. |
| $E(nw)$ | = | the average number of customers waiting in line in the queue, not including the one being served. |
| $E(tq)$ | = | the average time a customer spends in queue waiting for service and being served. |
| $E(ts)$ | = | the average service time per customer. |
| $E(tw)$ | = | the average time a customer spends waiting in line in the queue, not including the service time. |
| $m$ | = | the number of parallel servers in a multiple-server system. |
| $nq$ | = | the number of customers waiting in the queue, including the one being served. |
| $nw$ | = | the number of customers waiting in line in the queue, not including the one being served. |
| $Pb$ | = | the probability that all servers are busy in a multiple-server queue. |
| $P(nq = N)$ | = | the probability that the number of customers waiting in the queue is equal to the number $N$. |
| $P(nq \geq N)$ | = | the probability that the number of customers waiting in the queue is equal to or greater than the number $N$. |
| $P(tq > T)$ | = | the probability that the time a customer spends waiting for service and being served is greater than the time $T$. |
| $S(ts)$ | = | the standard deviation of the service times. |
| $tq$ | = | random variable characterizing the time a customer spends in the queue waiting for service and being served. |
| $ts$ | = | random variable characterizing the service time for a customer. |
| $tw$ | = | random variable characterizing the time a customer spends in the queue waiting for service. |
| $u$ | = | the utilization, or percentage of time busy, per server equal to $(a \times ts)/m$. |

to the analyst. Be prepared, then, to build a tailor-made generic data gathering software package.

When sufficient data has been gathered to provide confidence for statistical estimates, an estimate of the mean value and a standard deviation can be made in the traditional manner. Recall that the standard deviation of a random variable, $x$, is given by

$$S(x) = \sqrt{E(x^2) - E(x)^2} \tag{A-3}$$

In gathering such data, it is also often useful to note transaction types along with each measurement. Some transaction types naturally take longer than other transaction types, and a knowledge of their relative contributions to the overall average service time can be an important factor in system design considerations. For example, design alternatives, such as assigning different transaction types to different servers or using priority structures to improve throughput for specific transaction types, may wish to be considered in later option analysis. Without an appropriate breakdown of service time data by transaction type, the necessary insight for implementing useful options may not be available to the analyst.

Consider a system that handles three types of messages: administrative, data base inquiries (DBIs), and status messages. Suppose the mean service times for the three different transaction types had been measured, with the results shown in Table A-2.

The overall system mean service time is given by the weighted mean

$$E(ts) = \sum_{i}^{N} E(ts)_i * (P_i) \tag{A-4}$$

where $E(ts)_i$ = the mean service time of the transaction category and $P_i$ = the probability of occurrence of the ith category.

Thus, for all message types, the system mean service time is

$$E(ts) = .120 * \frac{84}{346} + .065 * \frac{160}{346} + .035 * \frac{102}{346} = .069 \tag{A-5}$$

**TABLE A-2 Calculation of Mean Service Time from Individual Service Times**

| Index (i) | Message Type | Mean Service Time (sec.) | Number of Occurrences |
|---|---|---|---|
| 1 | Administrative | .120 | 84 |
| 2 | DBI | .065 | 160 |
| 3 | Status | .035 | 102 |
| | | | N = 346 Total |

## The System Mean Arrival Rate

The next required statistic is an estimate of the mean arrival rate, $E(a)$, of customers requesting service. Arrival rate statistics are gathered by noting the time of arrival of each customer or transaction as it arrives at the system over a sufficient period of time to statistically gain confidence that your estimate is realistic. As with service time measurements, the arrival rates of consistent transaction types is also important to gather.

The analyst must be particularly sensitive to the occurrence of periodic cycles in arrival rates, such as hourly, daily, monthly, and annual fluctuations, and think about what he or she is doing. If arrival rates are high during afternoon hours and low during early morning hours, it will be of little value to use a 24-hour average arrival rate in the queueing analysis. In dealing with systems whose configuration is fixed, which is typically the case with automated systems, one will need to determine the arrival rate for transaction types at their busiest time interval, in order to design for the heaviest load. In dealing with queues that involve human servers, arrival rates should be determined at the busiest times, in order to design a configuration for peak loading, even though the system may be staffed at lower levels during off-hours.

Getting the right arrival rate pattern for the model may be simple or it may require considerable analysis. For example, queueing theory can be used quite effectively for determining staffing and equipment requirements in the public safety dispatching arena. Specifically, in Southern California fire departments, the arrival rate of calls for service at command and control centers in large cities can differ markedly from those encountered at county fire departments. Peak loads at city departments during July 3, 4, and 5 can be six to ten times the annual daily average. In county departments, the peak calls for service tend to exhibit a prolonged fluctuating increase during the months of the fall fire season. Often, designing for the required number of telephone, radio, and other command and control resources for absolute peak values can be financially prohibitive. In the county fire case, a reasonable approach may be to design to an average value of the peak values encountered. This so-called "peak average" is often used in systems where periodic sustained peak fluctuations occur. The city fire problem is more difficult because the peak is so acute relative to the rest of the year. An approach to this problem may be to design to, say, half the peak, but be prepared to assign additional resources, such as auxiliary radio channels and supervisor and training consoles, during the actual peaks. The point is that deciding how to gather and use arrival rate statistics is not simply a mechanical issue. Having gathered the data, the analyst must consider what that data means and develop a strategy for dealing intelligently with any observed fluctuations.

Statistical data gathering utility packages provided with computer systems seldom, if ever, include programs that gather arrival rate data. Again, be prepared to develop a custom-made program to tag, count, and time arrivals of requests for service at each computer subsystem or system resource at which it is anticipated that queueing analysis may be of value.

## ASSUMPTIONS AND LIMITATIONS

The following assumptions underlie the derivations of the queueing equations discussed in this chapter:

1. The number of customers is large compared to the number of servers in the system—that is, the customer population is equivalent to being mathematically infinite.
2. Customers do not leave the queue once they have entered the queue.
3. Arrivals for service are random (Poisson).
4. Service times are random (exponential).
5. Service is provided to customers on a first-in first-out basis.

Minor deviations from the above assumptions will always be encountered in reality. Generally, none of these are serious. If, however, the analyst is at all uncomfortable with general adherence to these assumptions, it may be wise to turn to simulation techniques. It does make sense, though, to consider applying analytic queueing at the outset if for no other reason but to bring focus to these issues.

A major limitation in using queueing theory occurs when system utilizations are high. As we discuss the basic equations presented in this appendix, it will be evident that the accuracy of the equations rapidly deteriorates as system utilizations approach values of 80 percent to 90 percent. The situation is mitigated somewhat in that systems with random arrival rates and service times are rarely designed to accommodate such high utilizations. In systems with reasonable utilizations, the queueing approach gives quite satisfactory answers.

Further caution must be taken in analyzing queues in tandem, as it is often the case that, while arrival rates may be random at the first queue, the transaction handling protocols in subsequent queues may affect this property of randomness down the line. Also, be particularly sensitive to the various ways that computer operating systems may impact the given assumptions in dealing with multiprocessing, paging, polling schemes, and the internal handling of files to storage media. For many of these issues, using simulation may be the better approach.

Despite these cautions, analytic queueing remains a powerful tool. Through thoughtful practice, the systems engineer will find it a valuable mechanism for rapid, top-level assessment of system performance and timely identification of performance bottlenecks.

## THE SINGLE-SERVER QUEUE

With a knowledge or good estimate of $E(ts)$, $S(ts)$, and $E(a)$, it is then possible to calculate a number of interesting properties of single-server queues. The standard single-server queue model is depicted in Figure A-1. In this paradigm, customers randomly arrive for service and wait, if necessary, in a single line for service from a single server. Customers experience an average waiting time, $E(tw)$, and, after waiting, an average service time, $E(ts)$. With their business done, they then leave the queue. The average time spent in the queue, then, is given by

$$E(tq) = E(tw) + E(ts) \tag{A-6}$$

The value for $E(ts)$ has been obtained from direct measurement. Determining the value for $E(tw)$ requires a knowledge of the average arrival rate, $E(a)$, which is also a measured value, and calculation of the system utilization.

The utilization, $u$, of a server in a queue is equal to the percentage of time that the server is busy conducting transactions. The average utilization over the time interval for which the average arrival rate was calculated is given by

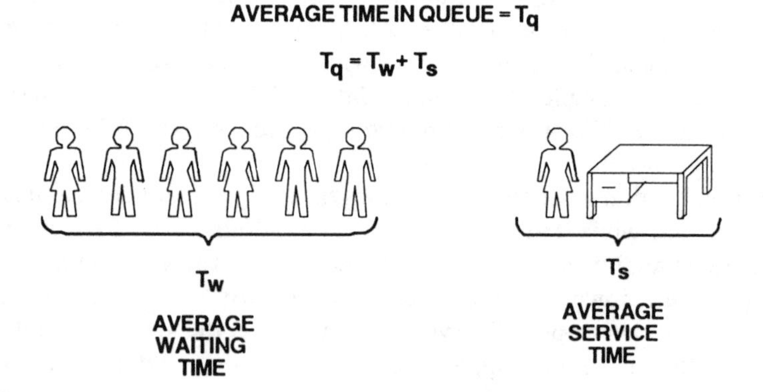


**FIGURE A-1.** The single-server queue.

$$u = E(a) * E(ts) \tag{A-7}$$

For example, if a queue experienced an arrival rate of 15 transactions per hour and the mean service time per transaction was 2 minutes, then the arrival rate per second would equal

$$E(a) = \frac{15}{3600} = .0042 \text{ per second}$$

and the mean service time would equal

$$E(ts) = 2 * 60 = 120 \text{ seconds}$$

and the utilization, or percentage of time busy, for the server would be

$$u = E(a) * E(ts) = .0042 * 120 = .50$$

or the server is busy 50 percent of the time on the average.

Note that the utilization, $u$, is dimensionless; thus the time units for $E(a)$ and $E(ts)$ must always be the same—in this case, the units were in seconds.

This example is somewhat intuitive. If the mean number of transactions per hour is 15 and the mean service time per transaction is 2 minutes, then the average time spent in service is 30 minutes (2 * 15), which is one-half, or .5, of the 1-hour time interval considered.

The next value of interest is the average waiting time for service ($Tw$) that a customer or transaction experiences waiting for service. This value is given by the Kinchene Polencheck equation

$$E(tw) = \left[\frac{u * E(ts)}{2(1-u)}\right] * \left[1 + \frac{S(ts)^2}{E(ts)^2}\right] \tag{A-8}$$

Examination of equation (A-8) shows that the standard deviation, $S(ts)$, of the service times has a crucial effect on the average waiting time $E(tw)$. Note that, if the standard deviation, $S(ts)$, is equal to the mean service time $E(ts)$, the right-most bracketed term of equation (A-8) is equal to 2. This value of 2 cancels out the 2 in the denominator in the left-hand portion of equation (A-8).

Alternatively, if $S(ts)$ is twice the value of $E(ts)$, then the right-most bracketed term becomes equal to 5 and the mean waiting time is increased by a factor of 5/2 over the case where $S(ts)$ is equal to $E(ts)$. On the other hand, if the value of $S(ts)$ is maintained at less than $E(ts)$, then the value of $S(ts)/E(ts)$ is less than unity and the value of the squared term works in favor of reduction, rather than rapid growth, of the resulting mean waiting time.

The sensitivity of this condition is portrayed in Figure A-2.

This reality provides the rationale for a basic rule when designing systems that involve service time statistics. The rule is that, if the standard deviation of the service times exceeds the mean value, it is wise to develop designs or modify procedures, such that the standard deviation is less than or equal to the mean.

In systems involving human transactions, this can be done by assigning longer transactions to a separate group (i.e, queue) such that the mean waiting times for shorter transactions are not made excessive by the occasional presence of long transactions. We would then have two sets of transactions—one with a shorter mean service time and one with a longer mean service time, but the grouping would be such that the standard deviation is less than or equal to the mean within each group.

For example, in the public safety radio dispatching environment (law enforcement, fire, and emergency medical services), a typical policy is to have one set of phone operators handle emergency calls and a second set handle

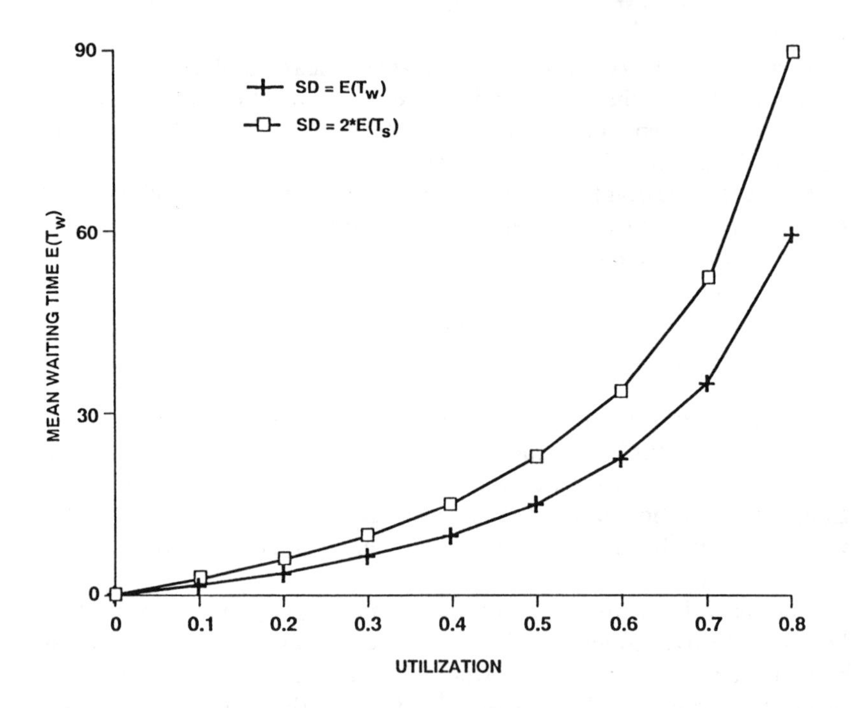


**FIGURE A-2.** Sensitivity of $E(T_w)$ to $SD$ ($SD$ = Standard Deviation).

longer administrative calls. This, of course, intuitively works well because we no longer have constant interruption of the longer administrative calls (preemptive interrupts) and/or longer average waiting times than desirable for emergency calls. The mathematical reason that this works (which is preferred to intuition) is that we no longer have a single queue where the standard deviation exceeds the mean service time.

In computer communications systems, a similar tack can be taken by assigning priorities to messages of identifiable types. The priorities are assigned such that, within each priority group, the standard deviation of the service times is less than or equal to the mean.

Another approach is to employ message blocking. In this scheme, message lengths are set at a length to accommodate most of the shorter messages. The long messages are then broken into the adapted smaller size, tagged with a message part number, transmitted, and reassembled at the destination according to their message number sequence. Each message then has a fixed length, and the standard deviation of service times becomes very small.

Both prioritization and message blocking techniques are designed to keep important and/or shorter messages from suffering extensive average throughputs at the expense of longer ones. The criteria for such designs is the relationship between the standard deviation and the mean service time of the observed service times in the system.

When the proper steps are taken to cause the standard deviation of service times to be less than or equal to the mean, equation (A-8) becomes simpler. That is, if the conditions are met such that $S(ts)$ is equal to or less than $E(ts)$, then it becomes conservative to represent equation (A-8) as equation (A-9), where $S(ts)$ is set equal to $E(ts)$

$$E(tw) = \frac{u * E(ts)}{(1 - u)} \tag{A-9}$$

Further, if the standard deviation becomes less than the mean, then using equation (A-9) as an estimate of the mean waiting time becomes even more conservative (e.g., yields estimates that are higher than may actually be encountered). It is also of mathematical interest that the exponential distribution has the property that the standard deviation is equal to the mean.

With the assumption that good design maintains the standard deviation of service times at a value less than or equal to the mean, we shall proceed using equation (A-9).

There are a number of additional properties of single-server queues with random (Poisson) arrival rates and random (exponential) service times that are useful to be able to estimate. These are given as follows.

The average number of customers, $E(nq)$, including those waiting and the one being served in a single-server queue, is given by

$$E(nq) = \frac{u}{(1 - u)} \tag{A-10}$$

The average number of customers, $E(nw)$, waiting in line to be served in a single-server queue, not including the one being served, is given by

$$E(nw) = \frac{u^2}{(1 - u)} \tag{A-11}$$

The average time spent in queue, $E(tq)$, including the waiting time of equation (A-9) and the service time, is given by

$$E(tq) = \frac{E(ts)}{(1 - u)} \tag{A-12}$$

The probability that the number of customers in queue, $nq$, including the ones waiting and the one being served, is equal to an integer $N$ is given by

$$P(nq = N) = (1 - u)\, u^N \tag{A-13}$$

The probability that the number of customers in queue, $nq$, including the ones waiting and the one being served, is greater than or equal to an integer $N$ is given by

$$P(nq \geq N) = \sum_{nq = 0}^{N} (1 - u)u^{nq} \tag{A-14}$$

The probability that the random variable, $tq$, characterizing the time a customer spends in a single-server queue waiting for service and being served is greater than a given time, $t$, is given by

$$P(tq \geq t) = 1 - e^{-(1-u)\, t/E(ts)} \tag{A-15}$$

Equations (A-10) through (A-15) can be used in a variety of useful ways. For example, Equations (A-13) and (A-14) are useful in estimating communications buffer sizes in computer systems, for limiting the probability of overflow to a desired level as a function of system utilization.

In human systems, equations (A-13) and (A-14) can be used for such issues as determining the number of chairs (increasing the number of chairs to match N) that should be in a physician's waiting room, such that the probability of no one having a seat is limited to a desired level. Working this problem requires a knowledge of the physician mean service time for the patient mix and a knowledge of how arrivals actually take place (which may or may not be similar to how they are scheduled). Adding more chairs, of course, increases the likelihood that everyone will have a seat, but that doesn't necessarily make everyone happier. Note that the waiting time for such poor souls does not change.

Equations (A-13) and (A-14), as with all of the equations given, are driven solely by the system utilization. Clearly, when a queueing problem exists, a utilization problem exists, and when utilization is high, there is either an arrival rate problem or a service time problem, or both. Thus, the physician inundated with complaints may wish to hire a colleague (i.e., create two single-server queues, each of which now has a lower arrival rate) or to review his or her patient scheduling algorithm (i.e., reduce the single-server arrival rate) or to find some way to treat his or her patients faster (i.e., reduce the mean service time). The extent and effect of each of these strategies can be predicted by using the single-server queueing equations.

### Example SS1

Consider a land line computerized data base system that receives incoming messages, supports a communications protocol, performs error detection and correction, performs a data base inquiry, stores messages, and then forwards them to a designated address. Further assume that a design requirement on the system is that the average message delay time at the node be less than 200 milliseconds.

The example uses a computer system on which this group of functions can be performed, with an overall system mean service time of 80 milliseconds and the standard deviation of the service times being less than this mean. Further assume that we wish to handle a peak load of 40,000 transactions per hour, or 11.1 transactions per second.

From equation (A-7), we estimate the system utilization at

$$u = E(a) * E(ts) = 11.1 * .08 = .89$$

Equation (A-9) yields a mean waiting time of

$$E(tw) = \frac{U * E(ts)}{(1 - u)} = \frac{.071}{0.11} = .645 \text{ seconds}$$

Equation (A-10) gives a mean number of transactions in queue as

$$E(nq) = \frac{u}{(1 - u)} = \frac{.89}{.11} = 8.0$$

Equation (A-6) gives an estimated total time in queue of

$$E(tq) = E(tw) + E(ts) = .645 + .080 = .725$$

Clearly, this design won't work, since the 725 milliseconds required exceeds our desired 200 millisecond average delay.

The problem is that the utilization is too high. This is always the problem. In dealing with single-server queues, there is a very simple point to reiterate. There are only two things that can be done to correct any single-server queue problem once the standard deviation of service times is already less than the mean. We must either reduce the arrival rate at a server or reduce the mean service time.

The arrival rate at a given server can be reduced by adding more servers, thus creating a parallel system of multiple single-servers. With this approach, the mean service time stays the same at each server. This method consists of duplicating the server system and dividing the load among the new group of servers. The utilization of each new server is reduced because the arrival rates at each are reduced proportionately. Hence, the mean waiting time is reduced. This is what supermarkets do when they open additional checkstands. They create multiple single-server queues. When human beings can observe the multiple single-server structure, the arrival rates at each server tend toward the same value, since the shortest lines are the ones joined first.

Alternatively, the mean service time can be reduced. There are many ways to reduce service times in computer systems. These include obtaining faster hardware (CPUs, disk units, display processors, coprocessors, busses, firmware, etc.) and/or obtaining or developing faster software operating systems, executives, and application algorithms.

Another method of reducing mean service times is to reduce the amount of work to be done by a given computer by distributing the load serially among two different computers. Of course, the response time for a given message now involves passage through two CPUs; however, because of the nonlinear property of equation (A-9), the attending reduction of service times in each machine compared with a single machine can potentially offset this.

In the current example, SS1, we shall assume the communications functions of supporting protocol and error detection/correction take approximately the same amount of time, on the average, as the storage function. We may thus be led to consider use of a front-end processor to handle communications and a second processor in series to handle data base inquiries. For simplicity, suppose the 80-millisecond service time is equally divided between the machines so that the mean service time on each will become 40 milliseconds. For each machine, then, the utilization would become

$$u = E(a) * E(ts) = 11.1 * .04 = .44$$

and the resulting mean waiting time estimate for each would become

$$E(tw) = \frac{(.44 * .04)}{(1 - .44)} = .031 \text{ seconds}$$

and the mean time in queue for each system would become

$$E(tq) = .031 + .04 = .071 \text{ seconds}$$

Since the machines are in series, the total mean queue time (waiting for service plus service) is twice the .071 seconds, or 142 milliseconds. We have thus been able to meet the 200-millisecond requirement by reducing the mean service times.

The example is admittedly simplistic in that the service times for both communications and data base functions were assumed to be equally divided among the original 80 milliseconds. The point, however, is not lost. Separate front-end processing is typical in communications systems. The approach is to divide and conquer by identifying two groups of basic functions whose service times are as equal as possible. This has the desirable potential of reducing utilizations to values, such that the exponential effect of the $(1 - U)$ term in the denominator of equation (A-9) does not drive the value of the mean waiting time, *Tw,* (hence the mean time in queue) to an excessive level.

Figure A-3 provides a feeling for the effect of system utilizations on system mean waiting times for single-server queues. Note that, when the utilization is equal to 0.5, the mean waiting time is equal to the mean service time. As system utilizations increase above 0.5, the exponential effect becomes more pronounced.

An acceptable value for system utilization is determined solely by the mean waiting time for service that can be tolerated by the system user. In systems where performance is critical, both humans and machines must generally experience lower mean utilizations—sometimes surprisingly low. This is a direct consequence of the fact that we are dealing with queues that experience random arrivals of requests for service and random service times.

If the arrival rates and service times were very nearly constant and the standard deviation of the service times was very small, then utilizations could approach values above 0.90. This might be the scenario with a machine designed to fill empty soda bottles as they move along a conveyor, come to a stop below a filling mechanism, and then move on.

In particular, note that, if the standard deviation of service times, $S(ts)$, is zero, then equation (A-8) reduces to the mean waiting time, $E(tw)$, for a single-server queue, with Poisson arrivals and with constant service times

$$E(tw) = \frac{u * E(ts)}{2(1 - u)} \tag{A-16}$$

In this case, the filling mechanism may experience a fairly high utilization. But when the arrival pattern is random, the effects of utilization on mean waiting times in single-server queues are those shown in Figure A-3.

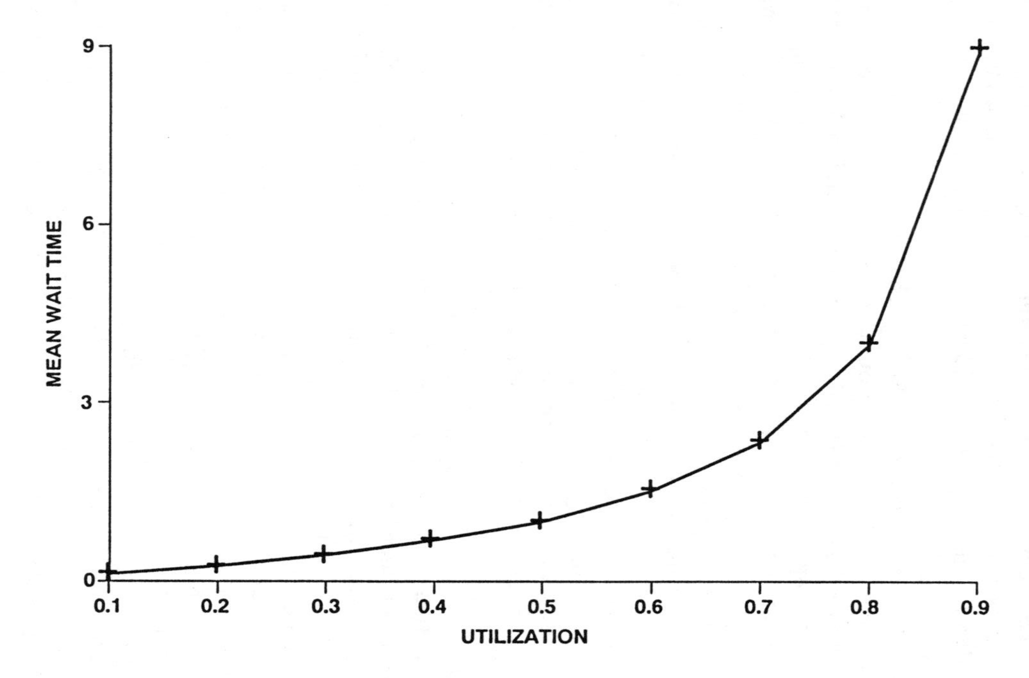


**FIGURE A-3.** Mean waiting time vs. utilization.

## THE MULTIPLE-SERVER QUEUE

The paradigm for the standard multiple-server queue is shown in Figure A-4. In this scheme, customers arrive for service at the queue and join a single line. The customer at the head of the line is serviced by the next available server. The multiple-server queue is commonly seen in banks and provides a higher throughput than a single-server queue, particularly at higher values of system utilization.

The equations of interest for the multiple-server queue are a bit more complicated, but far from unreasonable. The value for utilization, $u$, in the equations is maintained at a value of less than unity by dividing the product of the mean arrival rate, $E(a)$, times the mean service time, $E(ts)$, by the number of servers, $m$, in the system. This more general version of utilization is given by

$$u = \frac{E(a) * E(ts)}{m} \tag{A-17}$$

Note that, when $m$ is equal to 1, equation (A-17) is equal to equation (A-7) for the single-server queue. That is, when $m$ is equal to 1, the multiple-server reduces to a single-server.

The mean waiting time for the multiple-server queue, with $m$ identical servers, Poisson arrivals, and exponential service times is given by

$$E(tw) = [B/m] * \frac{E(ts)}{(1 - u)} \tag{A-18}$$

where $B$ is the probability that all $m$ servers are busy.

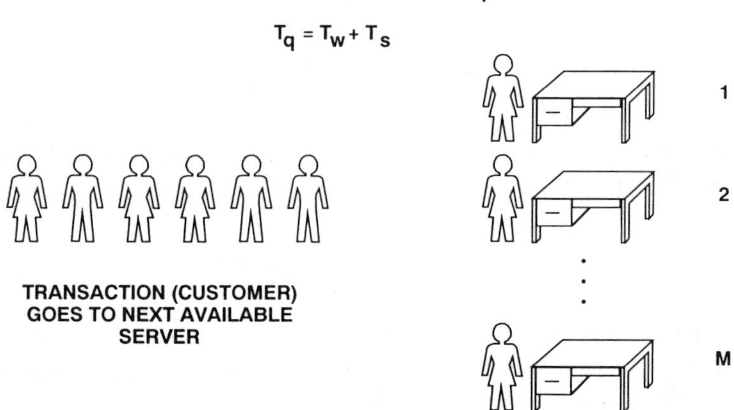


**FIGURE A-4.** The multiple-server queue.

The probability, $B$, that all $m$ servers are busy is given by

$$B = \frac{1 - \left[\frac{\alpha}{\chi}\right]}{1 - u\left[\frac{\alpha}{\chi}\right]}$$

where alpha equals

$$\alpha = \sum_{N=0}^{m-1} \frac{(m * u)^N}{N!}$$

and chi equals

$$\chi = \sum_{N=0}^{m} \frac{(m * u)^N}{N!}$$

Note that, when $m = 1$, the value for $B = u$ and equation (A-18) reduces to equation (A-9).

The following equations are also of interest in dealing with multiple-server queues.

The average number of customers, $E(nq)$, in a multiple-server queue, including those waiting to be served and those being served, is given by

$$E(nq) = B\left[\frac{u}{(1 - u)}\right] + m * u \tag{A-19}$$

The average number of customers, $E(nw)$, waiting in a multi-server queue, not including the ones being served, is given by

$$E(nw) = B\left[\frac{u}{(1 - u)}\right] \tag{A-20}$$

The average time, $E(tq)$, spent in a multiple-server queue, including the time waiting and the time being served, is given by

$$E(tq) = [B/m]\left[\frac{E(ts)}{(1 - u)}\right] + E(ts) \tag{A-21}$$

The probability that the waiting time random variable, *tw*, characterizing the time a customer spends in a multiple-server queue waiting for service is greater than a given time, *t*, is given by

$$P(tw > t) = Be^{-[m/E(ts)]\,[1 - u]\,t} \tag{A-22}$$

Multiple-server queues are commonly used in banks. They may also be used in computer communication systems, where two or more computers service incoming requests for information. One such application is an on-line electrocardiographic diagnostic aid system that provides analyses of electrocardiograms for physicians over telephone lines. In this system, incoming calls are directed to the first available machine among a bank of machines. Modern, large-scale parallel super-computers are also candidates for the use of multi-server strategies that make use of the next available processor.

Multiple-servers outperform multiple single-servers with the same number of servers. That is, consider the design alternative of implementing a single multiple-server system with two servers versus a system with two single-servers. The single multiple-server provides service to the customer at the head of the queue as soon as either one of the servers is not busy (the bank with two tellers). The ideal case with two single-servers is that the customers commit to one queue or the other, such that the number of customers in each queue is very near equal (the supermarket queue with two checkstands open). This equal loading is often the case when the customers can see the queues. It is also the case in computer systems that include algorithms to evenly distribute the incoming load among serving resources.

To gain a feeling for differences in performance, consider a system that experiences an average arrival rate, $E(a)$, of 12 customers per hour or 0.2 per minute and in which the mean service time, $E(ts)$, per server has been measured to be 3.0 minutes. We shall assume also that the standard deviation of the service times is less than the mean.

Table A-3 gives solutions for the mean waiting times, mean times in queue, and utilizations for a single-server queue with the full load, two single-server queues with the load equally divided among them, and a single multiple-server queue with two servers.

The example shows that two single-servers clearly outperforms one single-server, as you might expect. This improvement is due solely to the halving

**TABLE A-3 Comparison of Performance of Single-server and Multiple-server Queues (all entries are in minutes)**

| Queue | m | u | E(a) | E(tq) | E(ts) | E(tw) |
|---|---|---|---|---|---|---|
| Single-server | 1 | .6 | .2 | 7.5 | 3.0 | 4.5 |
| Two single-servers | 2 | .3 | .1 | 4.3 | 3.0 | 1.3 |
| Multiple-server | 2 | .3 | .2 | 3.3 | 3.0 | 0.3 |

of the arrival rate per server. More interestingly, the example shows that the mean waiting time, $E(tw)$, for the multiple-server queue is reduced by approximately one-quarter over that experienced with the dual single-server configuration. The overall queueing time, $E(tq)$, experienced by the customer is reduced by about 25 percent. This example was computed at a system utilization of 0.6. As the utilization increases, the relative performance of the multiple-server queue also increases.

An example of the use of queueing theory in conjunction with system reliability and availability considerations for establishing and assessing system architectures is provided in Chapter 12, "Trade-off Analysis."

**References**

1. Martin, J. 1967. *Design of Real-Time Computing Systems*. Englewood Cliffs, N.J.: Prentice Hall, Inc.
2. Kleinrock, L. 1976. *Queueing Systems*. Volumes 1 and 2, New York: John Wiley & Sons.

# Appendix B

# Reliability, Availability, Maintainability (RAM) Analysis

This appendix introduces a number of fundamental tools used for reliability, availability, and maintainability analysis. Discussion of how they are used to develop designs for system design trade-offs in concert with system support mechanisms is also included.

An example of how RAM analysis can be used in conjunction with queueing theory by the systems engineer to analyze and screen top-level system architectural concepts is given in Chapter 12, "Trade-off Analysis."

## RELIABILITY

A major requirement for conducting reliability analysis is an understanding of the failure rate of a unit. The term unit shall be used to stand for component, subassembly, subsystem, system, and so on. The failure rate, referred to as Lambda, is given by

$$\lambda = \frac{F}{T} \tag{B-1}$$

where $F$ is the number of unit failures over a given operating time, $T$.

There are three types of failure rates that are normally considered in reliability analysis. These are associated with periods of system life. The first is the period of *infant mortality,* and the second and third are the periods of *useful life* and *wearout*, respectively.

During the infant mortality phase, the failure rate is initially high and then declines until a relatively constant failure rate is experienced, which marks the beginning of the useful life phase. Major sources of failure associated with infant mortality are basically quality failures. These include manufacturing defects, design problems surfacing, and failures due to improper transportation and handling.

The useful life period is characterized by an assumed constant failure rate. During this period, failures are accepted to be random and are generally caused by physical, electrical, and environmental trauma.

During the wearout period, the failure rate begins to increase and continues to increase rapidly, until the unit is essentially no longer able to function. As the name implies, failures during this phase are largely due to age.

In our discussion, we will concern ourselves with the useful life period, which is usually the longest period in the life of a system. The reliability of a unit during the useful life phase is expected to behave exponentially, with a constant failure rate such that the probability that a given unit is still operational at the end of some period of time, *t*, is given by

$$R(t) = e^{-\lambda * t} \tag{B-2}$$

The reliability of a system that consists of *N* components in series is given by the product of the individual component reliabilities as follows

$$R_s(t) = \prod_{(i=1)}^{N} R_i(t) \tag{B-3}$$

where $R_i(t)$ is the reliability of the $i^{th}$ component.

The reliability of a system that consists of *N* components in parallel is given by

$$R_s(t) = 1 - \prod_{i=1}^{N} [1 - R_i(t)] \tag{B-4}$$

Equation B-4 assumes that all parallel components are active—that is, they are connected together and operating. This configuration is referred to as being redundant or exhibiting redundancy. In this scheme, only one of the components needs to remain operable in order for the entire parallel system to function.

If all *N* components in a redundant system have the same reliability (hence the same failure rate) then equation (B-4) reduces to equation (B-5)

$$R_s(t) = 1 - [1 - R(t)]^N \tag{B-5}$$

Equation (B-6) gives the formula for the system reliability of parallel components where only one of the components is active at a time. When a failure occurs, a switch is activated to cause another of the components in the parallel configuration to become active. This is referred to as standby redundancy. Equation (B-6) assumes perfect switching and an off-line failure rate of components that have not been used equal to zero (0).

$$R_s(t) = e^{-\lambda t}\left[\sum_{i=0}^{N-1} \frac{(\lambda t)^i}{i!}\right] \tag{B-6}$$

In this manner, a knowledge of the reliabilities for parts of systems, along with a knowledge of the system architecture, can be used to estimate reliability for the complete system.

For example, consider the redundant series parallel system shown in Figure B-1. Assume that each element in the system has an identical failure rate, lambda, equal to 1 failure in 10,000, or .0001. Suppose we wished to determine the system reliability at a time equal to 5,000 hours. The reliability for each of the elements under these conditions is given by equation (B-2) as

$$R(t) = e^{-.0001*5000} = .6065 = .61 \tag{B-7}$$

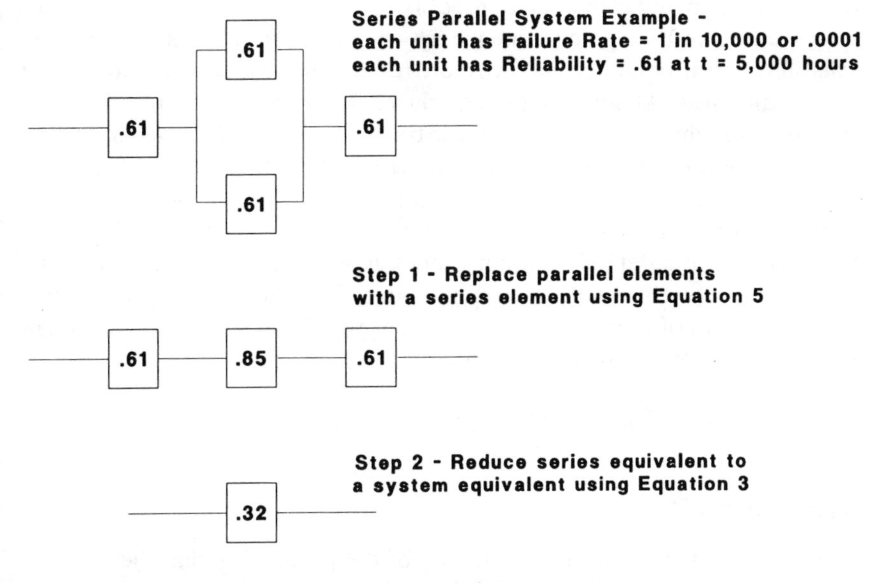


**FIGURE B-1.** System reliability sample calculation.

The first step is to replace the parallel elements with a single series representation. This is done by using equation (B-5)

$$R(t) = 1 - [1 - .61]^2 = .848 = .85 \tag{B-8}$$

We now have three elements in series with reliabilities of .61, .85, and .61, respectively. The next step is to reduce the series equivalent of the system to a single-system equivalent using equation (B-3)

$$R_s(t) = .61 * .85 * .61 = .32 \tag{B-9}$$

Thus, the probability that the system given in this example is still operating after 5,000 hours of use is .32.

It is also interesting to note the relative gains in placing elements in parallel. We have seen that two redundant elements in parallel, each of which has a reliability of .61, exhibit a reliability of .85 in parallel. This represents an improvement of a factor of approximately 1.4. If we use equation (B-5) again, to calculate the reliability of three such elements in parallel, we find

$$R(t) = 1 - [1 - .61]^3 = .94 \tag{B-10}$$

Note that, while the reliability of the configuration has increased to .94 (considerably better than the original .61), the factor of improvement in going from two redundant elements to three is only 1.1, as opposed to the factor of improvement of 1.4 in going from one to two redundant elements. Thus, while reliability will always improve with more redundancy, the relative returns will always diminish. At some point, clearly, element costs come to grips with reliability requirements. More on these trade-offs will be discussed following treatment of the *mean time between failure* (MTBF) and system availability.

Some systems may be considered to have met their reliability requirements if $m$ out of $N$ specific units are operational. One example of this might be a shared computer information system, in which it is deemed acceptable to have at least eight out of ten terminals operating. The reliability of a system in which $m$ out of $N$ identical units, each of which have reliability $R(t)$, are required to be operating is given by

$$R_s(t) = \sum_{k=m}^{N} \binom{N}{k} R(t)^k [1 - R(t)]^{N-k} \tag{B-11}$$

## AVAILABILITY

The availability of a unit is an estimate of the probability that the unit can actually be used for its intended purpose in its operational environment at any point in time. It is a function of unit reliability (MTBF) and the *mean downtime* (MDT) for the unit. Downtime can be due to a number of reasons, including unit failure, necessitating a repair or replacement time, planned maintenance time, time involved in the logistics of acquiring needed parts or service personnel, or by administrative time delays.

Availability can be viewed as the ratio of useful time to total time over a given time period, where the total time includes both useful time and non-useful time. This ratio for availability, $A$, is generally expressed using the MTBF and the MDT as

$$A = \frac{MTFB}{[MTBF + MDT]} \tag{B-12}$$

The systems engineer may assign activities to the MDT as he or she sees fit for the need. Three standard means of making such assignments are given in Table B-1.

Inherent availability is a useful concept in that it provides a direct measure of the performance of the repair function alone. Note that the MTTR should logically include the time to isolate a failure, the time to actually make a replacement, and the time to verify that the system is once again functioning before returning to an operational state. Achieved availability provides a more realistic measure of actual availability of the system to the user in systems where planned maintenance is a matter of routine policy.

Operational availability includes the characteristics of inherent and achieved availability plus logistics factors. For example, if a vendor's customer engineer is not on site at the time of a failure, travel time will clearly need to be added into the system restoration time. Nor do most systems carry complete inventories of parts on site, so that the time for parts acquisition after fault isolation is often a realistic consideration in determining true system availability.

The term mean time to restore is often used in conjunction with operational availability to distinguish between the simpler consideration of repair time alone (mean time to repair) and the overall time to actually restore the system to operations (mean time to restore), which includes logistics. Unfor-

**TABLE B-1 Three Standard Availability Measures**

| Measure | Definition |
|---|---|
| Inherent | MDT = mean time to repair (MTTR) |
| Achieved | MDT = MTTR plus planned maintenance |
| Operational | MDT = MTTR plus planned maintenance, plus logistics (parts, administrative, supplies, etc.) |

tunately, the words repair and restore both start with the letter R so that the term MTTR is sometimes ambiguous. In this writing, the more generic term, mean downtime (MDT), is used to represent any of the various activities that may sum up to a period in which the system is considered to be unavailable to the user. The systems engineer will want to give some thought to the definition of MDT as a function of how the system is used and/or supported. In some cases, using more than one measure may clearly be appropriate in order to understand different aspects of performance of the support mechanism.

## MEAN TIME BETWEEN FAILURE

In addition to the MDT, an estimate of the MTBF for each unit is required for determinating unit and system availabilities. The MTBF for a unit is an estimate of the expected average time duration between failures for that unit. The MTBF for the useful period is given by

$$MTBF = \int_0^\infty R(t)dt = \int_0^\infty e^{-\lambda t}dt = \frac{1}{\lambda} \tag{B-13}$$

Thus, the MTBF during the useful period of life is just equal to the reciprocal of the failure rate.

The MTBF for $N$ units in series is given by

$$MTBF_s = \frac{1}{\sum_{i=1}^{N} \frac{1}{MTBF_i}} \tag{B-14}$$

where $MTBF_i$ is the MTBF of the $i^{th}$ unit of $N$ units.

The MTBF for $N$ identical units in parallel is given by

$$MTBF_s = MTBF_c \sum_{i=1}^{N} \frac{1}{i} \tag{B-15}$$

where the MTBF for each of the $N$ components, or units, is $MTBF_c$ and there is no replacement until *all* units fail.

When units are placed in parallel, a dramatic improvement in the MTBF for the parallel configuration (hence, availability) can be realized if any unit that fails is restored to service immediately upon failure rather than waiting for all units to fail. The MTBF for two parallel identical units under the condition that either unit is restored upon failure is given by

$$MTBF_s = ½\left[MTBF_c\,(3 + \frac{MTBF_c}{MDT_c})\right] \tag{B-16}$$

For example, consider a unit with a failure rate of .001 (MTBF = 1,000 hours) and an MDT of 100 hours. The unit availability is given by equation (B-12) as

$$A = \frac{MTBF}{(MTBF + MDT)} = \frac{1{,}000}{(1{,}000 + 100)} = .91 \tag{B-17}$$

The equivalent MTBF for two units in parallel without replacement is given by equation (B-15) as

$$MTBF_s = MTBF_c * 3/2 = 1000 * 3/2 = 1{,}500/\text{hours} \tag{B-18}$$

Again, by equation (B-12), the availability for the two parallel units without replacement becomes .94.

The equivalent MTBF for the parallel configuration with immediate replacement upon any failure is given by equation (B-16) as

$$MTBF_s = ½\left[1000 * (3 + \frac{1000}{100})\right] = 6500 \text{ hours} \tag{B-19}$$

With this equivalent MTBF, equation (B-12) yields a system availability of .98. The MDT used in this example of 100 hours is admittedly in excess of what is normally encountered. The values were selected to emphasize the relative gains that may be achieved in availability via the alternative strategies.

Equations 15 or 16 can thus be used to reduce the MTBFs for parallel units to one MTBF value for a single equivalent series unit. In this manner, a system consisting of a number of series and parallel units can be reduced to an equivalent representation that consists solely of units in series. Equation (B-14) can then be used to determine the MTBF for the complete system, $MTBF_s$.

The system equivalent mean downtime, $MDT_s$, for a number of units in series is given by

$$MDT_s = MTBF_s \sum_{i=1}^{N} \frac{MDT_i}{MTBF_i} \tag{B-20}$$

Equation (B-12) can then be used to determine the system availability for a complete system, $A_s$,

$$A_s = \frac{MTBF_s}{[MTBF_s + MDT_s]} \tag{B-21}$$

Also note that substitution of equations (B-14) and (B-20) into equation (B-21) yields an alternative means of determining the system availability, $A_s$, for a system of $N$ serial units

$$A_s = \frac{1}{1 + \sum_{i=1}^{N} \frac{MDT_i}{MTBF_i}} \tag{B-22}$$

Consider the example illustrated in Figure B-2. A system consists of four units, two of which are in series with a parallel combination. Unit 1 has an MTBF of 10,000 hours and an MDT of 18 hours. The two parallel units, designated as unit 2 in the figure, each have an MTBF of 2,000 hours and an MDT of 48 hours. The last serial unit, unit 3, has an MTBF of 8,000 hours and an MDT of 36 hours.

The first step is to replace the parallel units with a single series unit. Using the non-replacement strategy of equation (B-15) results in a single series equivalent unit, 2E, with an MTBF of 3,000 hours. The MDT is assumed to remain at 48 hours. The resulting series system is then reduced to a single-system equivalent using equation (B-22), resulting in a system availability, $A_s$, of .978.

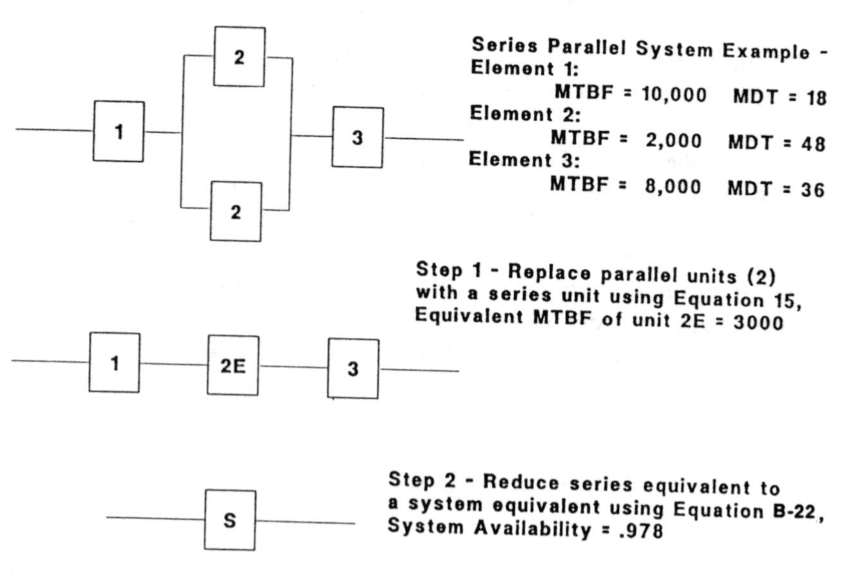


**FIGURE B-2.** System availability sample calculation.

It is worth noting that the equivalent MTBF of the parallel units in the example using a strategy of immediate replacement upon the failure of either unit (equation (B-16)) would yield an equivalent MTBF of 44,667 hours. This represents a decided improvement and clearly suggests that the parallel replacement strategy can essentially eliminate the undesirable effects of low-reliability units in a series system.

The block diagram that constitutes the reliability and/or availability models, of course, are different from the normal functional system block diagram. In system reliability/availability models, all units required to achieve a specific functionality appear in series. This is because, if one needed unit fails, the entire system fails to function. For example, environmental control units (heaters, air conditioners, etc.) may not always appear in a functional diagram, but must be included as series elements in a reliability/availability model if equipment operation is dependent on the maintenance of the environment. Certainly, some of the units may employ redundancy to increase reliability, but the redundant units still must appear in series mathematically.

It is also important to consider that a given reliability/availability model may not include all units of the entire system. For example, the ability of a system to produce reports may be of less importance than the ability of the system to receive, process, and forward messages. Figure B-3 illustrates this point. In this example, the reporting and message handling functions employ different units of the system (hence, the determination of the availability of these functions involves two different models). Each of these models will include only those units of the system required to accomplish its function.

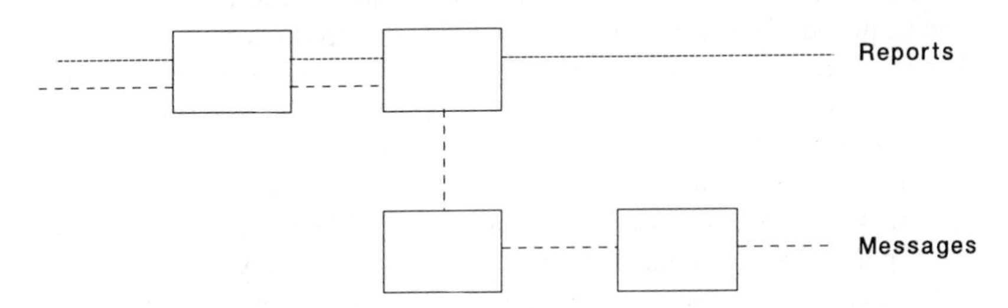


In general, system availability is not determined by lumping all components together in one equation.

Availability should be determined for each functional path.

Typically, the paths are different and the requirements for each functional path are different.

**FIGURE B-3.** Functional availability paths.

Similarly, entire systems, such as aircraft, may have different availability requirements for different missions.

Typically, the availability requirements for different functions of a system are also different. Thus, a given system may have any number of reliability/availability models, depending on the functional requirements of the system. Perhaps one of the most important roles of the systems engineer in RAM analysis is to take the lead in defining functional paths that need to be included in system RAM analysis. These considerations should be addressed at the time of setting system reliability and availability requirements, which should be included as a part of the system functional requirements.

The analytic techniques and strategies presented in this appendix, along with elements of queueing theory treated in Appendix A, are employed in Chapter 12 to illustrate an effective approach to determining and/or validating system architectural concepts.

## PREDICTING MTBF AND MDT

As is frequently the case in building models, acquiring good data can be the most difficult part. In RAM analysis, the problem focuses on gathering accurate MTBF and MDT data. Like many of the activities in systems engineering, a detailed understanding of specific values is gained only through time as designs mature and opportunities for adequate testing emerge. Often, actual values for MTBFs can be gained only empirically, after systems have been in operation for a statistically significant period of time. Prior to the time at which an accurate understanding of MTBFs and MDTs can be acquired, the systems engineer must make predictions.

Throughout the requirements definition, preliminary design, and detailed design phases, the systems engineer must constantly track the uncertainty associated with different units of the RAM model. If the uncertainty is discomforting or if the issue is of critical importance to system performance, the progress of acquiring good RAM figures should be routinely tracked as a formal part of margin management. As systems engineer, you may also wish to include representation on your design team for this element of specialty engineering through routine involvement of a RAM expert through any critical periods. The RAM expert will not only assist in validating the team's basic approach to analysis, but should also support the development of a realistic program to methodically reduce uncertainty through testing and risk management techniques.

It should also be clearly understood that the ability of the system to meet its functional reliability requirements cannot truly be measured until all such MTBF data is validated. Similarly, the ability of a system to meet its functional availability requirements cannot be accurately measured until both the MTBF and MDT data are validated.

There are basically three ways to predict MTBF. The easiest occurs when utilizing previously used equipment or equipments that are very similar, where empirical data is available. Probably the highest measure of a vendor's cooperation is their willingness to provide MTBF data on their products. Many simply will not, but sometimes the information can be gleaned with appropriate cajoling.

The second situation involves estimating MTBFs for equipment that may be functionally different from existing products but that involve very similar components. In this case, one may build their own subsystem reliability/availability model, based on a knowledge of MTBFs for units included in the subsystem. It is also useful to seek out experts in the appropriate disciplines outside a given project or even outside the organization itself.

The last and most difficult condition arises when dealing with totally newly designed equipment. In this case, analytic derivation of failure rates for individual components may be necessary to estimate subsystem MTBFs until actual data can be acquired. Electronic part failure rate prediction techniques are available in the current edition of MIL-HDBK-217C. While these analytic techniques tend to be conservative, they at least provide a starting point when no other recourse is available.

The most difficult part of MDT prediction lies in estimating the MTTR. One standard, given in MIL-HDBK-472, provides a predictive approach for the average MTTR that can be expressed as

$$E(MTTR) = \sum_{i=1}^{N} \frac{[\lambda_i * MTTR_i]}{\sum_{i=1}^{N} \lambda_i} \tag{B-23}$$

where lambda$_i$ = predicted failure rate of the i$^{th}$ element;
$MTTR_i$ = MTTR of the i$^{th}$ element; and
$N$ = the number of units.

MDT predictions are dependent upon the design of the system support mechanism. Trade-offs between on-site repair and restoration versus modular replacement of failed units with off-site repair are discussed in the next section.

## SUPPORT MECHANISM LOGISTICS

There are basically only two things that can be done to improve system availability. The first is to increase the MTBF of system units, and the second is to decrease the system MDT. As we have seen, MTBF can be increased by employing units with higher reliability and/or employing unit redundancy.

Decreasing achieved and operational availability are achieved by streamlining maintenance procedures and by reducing the time associated with operational logistics. Decreasing inherent availability must concentrate on decreasing MTTR. Decreasing MTTR must, in turn, concentrate on reducing time allocated for fault isolation, repair or replacement time, and time for validation of the repair.

Among the most efficient maintainability designs to accomplish the reduction of MTTR is commonly found in military systems. Such critical systems employ built-in test equipment (BITE), which performs built-in tests (BIT) on modularized subsystems or portions of subsystems. The BIT is either initiated on an automated schedule or can be initiated on demand prior to a specific mission. The function of a BIT is to reduce the time for fault isolation.

The purpose of partitioning the system into modules is to reduce the time to restore. Modularization also reduces the complexity of the BITE. The fault isolation is performed on subsystems or parts of subsystems, such that they can easily be swapped out by a single member of the maintenance team. A simple Go or No-Go, green or red light, indication is employed, such that a sophisticated technical knowledge is not required to rapidly accomplish repair. Reinitiation of the BITE is then used to validate that the repair by replacement has been successful and that the system has been restored to an operational state. Modules treated in this fashion are often referred to as *field replaceable units* (FRUs). In this manner, the functions of fault isolation, repair, and validation of restoration can be carried out in a minimum of time. Typically, the remainder of the support mechanism is operationally off line. Modular units that have failed BIT are removed from the operational scene to intermediate areas, where screening for part replacement and simple repair takes place. At this level, modules may be taken apart and units of the modules may be tested and replaced as required. These module units are referred to as *shop replaceable units* (SRUs). If serious problems are encountered beyond the SRU level that entail further analysis and repair, the FRUs or the SRUs are returned to a rear area that is supported by more sophisticated repair capability and by vendors.

Note that the basic philosophy of this support mechanism involves the timely replacement or movement of equipment at the modular level among the various support entities. In such systems, it is a matter of considerable interest to be able to estimate the number of standby, or spare, FRUs and SRUs that must be in hand at the appropriate echelon to maintain the flow demand between each echelon. That is, when front line maintenance personnel send a defective module to an intermediate area for repair, they don't really care if they ever see that particular module again. They only care that they have enough modules on hand to support their needs at any given time. The flow rate of equipment between echelons is determined by the number of spares required in inventory at each echelon.

Equation (B-24) provides a useful means for calculating the number of spare units required for any system in which failure rates and the logistics time for unit supply can be estimated. The probability, $P(R)$, that a number of units less than or equal to $R$ fail in a given time period, $t$, is given by

$$P(R) = \sum_{N=0}^{R} \left[ \frac{(\lambda * t)^N}{N!} \right] e^{-\lambda * t} \qquad \text{(B-24)}$$

where $N$ is the number of units in the system, with a failure rate equal to lambda.

For example, consider a system that employs ten terminals, where each has an MTBF of 8,000 hours (or a failure rate of .000125). Assume that the vendor can provide you with terminals 10 days, or 240 hours, after you order them. Under these conditions, equation (B-24) estimates that the probability of less than or equal to one unit failing in 240 hours is .963. Conversely, the probability that more than one unit fails in 240 hours is just given by $1 - P(R)$, which is .037. Thus, there is about a 96-percent confidence level that only one spare terminal is required.

A further iteration of equation (B-24) shows that the probability of less than or equal to two units failing in 240 hours is estimated at .996. The probability of more than two units failing is .004. Thus, if your policy dictated that you wished to be 99-percent sure that you would not run out of spare units, you would want to keep two units in inventory for the values presented in the example. The simplest way to use equation (B-24) is to write a small iterative program in which you set a desired confidence level. The program starts with $R = 1$ and then iterates on the value $R = R + 1$ until the desired value for $P(R)$ is met or is exceeded. The value for $R$ then yields the number of spares required in inventory.

In less stringent government and commercial systems, the support mechanism will characteristically employ only two levels consisting of on-site support and external vendor support. In many large systems, vendor support is permanently on site in the form of customer engineers (CEs). Such CEs have permanent offices at a preferred customer's base of operations and use this base to service the main customer as well as other customers in the immediate area. If you are involved in a system that could fit this mold, it is worth looking into the vendor's policy, as the logistics factors of the MDT equation are consistently reduced in support of operational availability when the CE is at your site.

The simplest maintenance scenario entails vendor support on a call-as-needed basis, usually supported by a maintenance contract. When system availability is not critical, this can be a satisfactory arrangement. You need not, however, restrict yourself to a single strategy for all time. For example, a ground support system for a spacecraft during a planetary encounter may employ a combination of increased spare unit inventories, along with intense on-site vendor support during this critical period, and then revert to a less demanding support mechanism during less critical cruise operations. The effects on system availability of such strategies can be estimated through the attending modification of MDT on each of the units affected.

Vendors, of course, do not commonly allow users to attempt to repair their equipment, and often stipulate this condition in maintenance contracts. In these cases, MDT can be significantly reduced by maintaining spares on site, such as spare terminals, cables, keyboards, printers, and CRTs. On-site personnel belonging to the user organization can swap such units in and out, without violating contractual arrangements. Equation (B-24) can be used to determine the number of units required for inventory.

Mainframes are usually too expensive to maintain idle backup spares. Many systems, however, employ one mainframe for operations and a second for software maintenance and continued development. With this arrangement, the development machine can serve as an operational backup, thus greatly reducing system MDT (hence, effective system availability).

It is evident that the designer of the maintainability component of RAM has many options. The basic task is to carry out a balanced RAM analysis in the context of the complete Life Cycle Cost setting—admittedly, no small task. Like all such analyses, however, the work should first be carried out at a rudimentary top level to gain overall insight into potential problem areas. The analysis is then extended in detail as required and as the design matures.

The starting point is to realistically set the system availability functional requirement. This requirement must be determined by the system user within the SDT setting.

The next step is to perform a RAM analysis on the functional system architecture that the design team has agreed upon, to meet functional performance requirements. The functional architecture does not have to be absolutely finalized. It may change as the design process proceeds, but it is better to start as early as possible. On the other hand, it is not usually productive to start your RAM analysis on some arbitrary architecture that is not driven by an availability requirement.

If meeting the system availability requirement on the performance-oriented architecture is difficult (it may not be), then MTTR and/or MDT issues need to be addressed. Work on the weakest link first, using combinations of higher-reliability parts, redundancy, and MDT reduction concepts until the next weakest link becomes evident. Then work on that one. Work in this balanced fashion until the system availability requirement is met.

You must also make top-level cost estimates for each alternative considered. Employing higher reliability units, redundant units, and strategies to reduce MDT all cost money. Both development and operational costs for each proposed alternative need to be estimated. BIT, for instance, increases development costs significantly, but can reduce operational costs to the extent that life cycle costs are actually lower.[1] As a guide to considering MDT costs, review the basic elements of logistics support (Chapter 11), which include supply support, required test equipment, transportation and handling, technical documentation, facilities, personnel, and training.

When a RAM strategy for meeting the system availability requirement has been arrived at and attending costs have been estimated, there may be a conflict between perceived cost guidelines and system availability requirements. As systems engineer, you then turn to your PCDCs to determine the best course of action (see Chapter 12). The previously agreed upon PCDCs should provide a definite guideline as to which characteristic must be subordinate (within limits). If you feel that the issue is beyond the normal province of the SDT's decision-making arena, raise the issue with your management. Present the problem *along with* the design team's user-oriented recommendations.

[1] The trade-off is made even more interesting when the development and operational phases are funded by two different entities, each of which wishes to minimize costs.

# Index

A000022416483